Philippine Digital Cultures

Asian Visual Cultures

This series focuses on visual cultures that are produced, distributed and consumed in Asia and by Asian communities worldwide. Visual cultures have been implicated in creative policies of the state and in global cultural networks (such as the art world, film festivals and the Internet), particularly since the emergence of digital technologies. Asia is home to some of the major film, television and video industries in the world, while Asian contemporary artists are selling their works for record prices at the international art markets. Visual communication and innovation is also thriving in transnational networks and communities at the grass-roots level. Asian Visual Cultures seeks to explore how the texts and contexts of Asian visual cultures shape, express and negotiate new forms of creativity, subjectivity and cultural politics. It specifically aims to probe into the political, commercial and digital contexts in which visual cultures emerge and circulate, and to trace the potential of these cultures for political or social critique. It welcomes scholarly monographs and edited volumes in English by both established and early-career researchers.

Series Editors
Jeroen de Kloet, University of Amsterdam, The Netherlands
Edwin Jurriëns, The University of Melbourne, Australia

Editorial Board
Gaik Cheng Khoo, University of Nottingham, United Kingdom
Helen Hok-Sze Leung, Simon Fraser University, Canada
Larissa Hjorth, RMIT University, Melbourne, Australia
Amanda Rath, Goethe University, Frankfurt, Germany
Anthony Fung, Chinese University of Hong Kong
Lotte Hoek, Edinburgh University, United Kingdom
Yoshitaka Mori, Tokyo National University of Fine Arts and Music, Japan

Philippine Digital Cultures

Brokerage Dynamics on YouTube

Cheryll Ruth Soriano and
Earvin Charles Cabalquinto

Amsterdam University Press

Cover image: Maysa Arabit

Cover design: Coördesign, Leiden
Lay-out: Crius Group, Hulshout

ISBN   978 94 6372 244 5
e-ISBN   978 90 4855 244 3 (pdf)
DOI   10.5117/9789463722445
NUR   740

Printed and bound by CPI Group (UK) Ltd, Croydon, CR0 4YY

Table of Contents

List of Tables and Figures

Acknowledgements

We could not have completed this project without the individuals, groups and institutions that have contributed to the birth and development of this project, and accompanied us in our respective academic journeys.

We thank Catherine Gomes for the encouragement and for referring our proposal to the series on Asian Visual Cultures of Amsterdam University Press – the nudge that started it all. Earlier ideas that became the foundation for Chapters four and five were presented at the Association of Internet Researchers 2019 conference. We thank Jonathon Hutchinson and Mary Elizabeth Luka, Series editors of the Special Issue, 'Trust in the System,' for Information, Communication and Society where our study on sisterhood solidarities on YouTube had been published. The reviews helped inform the data analysis and broader ethical considerations for this book. Our article on YouTube and platform labor influencers was published in *Sociologias* and we thank Rafael Grohmann, Ludmila Abilio, and Henrique Amorim for the feedback for that piece that in turn helped shape chapter Five. Some of the key ideas in Chapter Six were presented at the Democracy and Disinformation national conference in February 2021. It helped us reflect on the ways YouTube and influencer dynamics have been reconfiguring Philippine politics along with its broader implications – and shaped some key arguments of the Concluding chapter.

Our thanks go to Maryse Elliot and Jasmijn Zondervan and the Amsterdam University Press Editorial Team for their editorial guidance and supervision. We also thank Fatima Gaw, Joy Panaligan, Marge Medina, and Arvin Mangohig for admirable diligence as research assistants for this book.

Our special thanks to the reviewers for their valuable comments and suggestions that helped us not only in sharpening our arguments but in appreciating the broader contributions of the book to digital cultures and Philippine studies scholarship. We are grateful to our universities – De La Salle University (DLSU) and Deakin University for providing institutional and financial support.

From Cheryll—
The assurance of love and support from family, friends, and colleagues helped me power through the writing of the manuscript in the midst of a pandemic. My heartfelt thanks to my family—Pacita, Darwin, and Chanel Nicole Soriano, and my partner, Tony Lopez----their love and daily acts of generosity supported me in many ways throughout my academic life.

DLSU provided a book writing grant that allowed me to carve out valuable time to write this manuscript. Colleagues in the Department of

Communication were a constant source of cooperation and intellectual stimulation. My exchanges with colleagues and students about everyday digital cultures and the function of YouTube in their lives informed several ideas in this book.

I also appreciate the many mentors and research collaborators who have supported my work over the years and the scholars whose work helped me think critically about the multiple articulations of power in media, platforms, power, and cultural production: Sun Sun Lim, TT Sreekumar, Millie Rivera, Jace Cabanes, Jan Bernadas, Adrian Athique, Heather Horst, Emma Baulch, Anjo Lorenzana, Gerard Goggin, Raul Pertierra, Anna Pertierra, Larissa Hjorth, Mohan Dutta, Jack Qiu, Julie Chen, Fernanda Pires de Sa, John Nery, Jonathan Ong, Lia Uy-Tioco, among others.

I'm thankful for the opportunity to present the theoretical backbone of the book and obtain critical feedback from social science and Philippine studies experts at the PhilS4 series organized by Sydney SE Asia Centre, UP Diliman, Humboldt University, and SOAS University of London.

From Earvin--
I would like to thank the people who are part of my book writing journey. While away from my family in the Philippines, I received support from my father Felix B. Cabalquinto, my siblings, and their families. I am also grateful to my partner, Guy, and his family, for providing the respite from book writing.

I express my appreciation to my caring friends. In Melbourne, I had generative and critical conversations with Oscar Serquina, Laurence Marvin Castillo, and Katrina Ross Tan. I enjoyed writing breaks with Eden Tongson, Jappy Alana, Nina Araneta-Alana, Alex dela Cruz, Eunice Guzon, and Cheenee Otarra-Garde. I had energising chats with my friends in the Philippines, including Eliza de la Fuente, Ces Nepomuceno, Elmirah Salanga, and April Castillo.

My foray into critically unpacking digital cultures and content creation has been inspired by my interactions with scholars in the field. When I participated in the 2018 Digital Media Research Centre Summer School, I was motivated by my conversation with Professor Jean Burgess to investigate the impacts of YouTube in a Global South context. I have also been privileged to engage and learn with passionate friends in academia, including Akane Kanai, Benjamin Hanckel, Natalie Hendry, Brady Robards, Annisa Beta, Thomas Baudinette, Jenni Hagedorn, Kate Mannell, Indigo Holcombe-James, Jian Xu, Monika Winarnita, Emily van der Nagel, Cesar Albarran-Torres, Alexia Maddox, Koen Leurs, and Andy Zhao.

To the almighty, and to my mother in heaven, thank you for always bringing inspiration, joy and the strength to keep going.

1 Lights, Camera, and Click the Notification Bell!

Abstract

This introductory chapter underscores YouTube as a critical site for understanding the rapid social, economic, political, and digital transformations in neoliberal and postcolonial Philippines. It situates brokering on YouTube within the evolving political, economic and media systems in the Philippine context, foregrounding the distinctive affective performances of Filipino YouTubers in networked publics. This chapter also presents the methodological considerations of the research study as well as the organisation of the book.

Keywords: brokerage, digital cultures, neoliberal globalisation, social media, postcolonial Philippines, YouTube

In recent years, we have witnessed the meteoric rise of influencers. Ordinary individuals expose their most intimate lives, generating massive followers. YouTube, as one of the most popular online platforms in the twenty-first century, has birthed influencers, including Swedish YouTuber PewDiePie and YouTube's youngest millionaire Ryan Kaji as examples. In the Philippines, YouTubers such as the late Lloyd Cadena, Michelle Dy, Zeinab Harake, Ja Mil, Ivana Alawi, and Mimiyuuuh, to name a few, have occupied the platform, with their channels becoming embedded in the Filipinos' everyday lives. These figures have essentially become constitutive of Filipinos' appetite for the consumption of informative and entertaining content. Whether based in or outside the Philippines, they are already part and parcel of Filipino households. Notably, their creative take on genres, unique performances, and interactive branding strategies are paving the way for their visibility, popularity, and even amassing fortune within and beyond social media. In a networked space, they produce, distribute, and monetise mundane, intimate, and random content, ranging from exclusive

Soriano, Cheryll Ruth and Earvin Charles Cabalquinto: *Philippine Digital Cultures: Brokerage Dynamics on YouTube*. Amsterdam: Amsterdam University Press, 2022
DOI: 10.5117/9789463722445_CH01

confessions, skills and knowledge enhancement, talent showcase, up to as outrageous as playing a prank on someone. At the other end of the screen, diverse audiences watch, like, or share these contents and can opt to subscribe or follow content creators that resonate with them. It is through this dynamic and networked environment where sociality, aspirations, and profit-making coalesce.

Of late, YouTube's role in the media ecosystem has expanded from catering to lifestyle videos into being somewhat of an "academy," where we find a broad range of information that build skills and exchange know-how, or what Utz and Wolfers (2020) call "epistemic communities." YouTube has turned into an interactive and shared space where creators share information and experience while users watch and learn in a social environment. This development collides with the participatory and do-it-yourself (DIY) culture in social media (Jenkins, 2006) and the cultural norms surrounding the "broadcast yourself" culture on YouTube, which facilitated an environment conducive to new modes of discovery and learning from the ordinary person, the experts of lived experience. This use of the academy metaphor for YouTube aligns with the claims of Susan Wojcicki, YouTube's Chief Executive Officer, indicating that YouTube is "more like a library in many ways, because of the sheer amount of video that we have, and the ability for people to learn and to look up any kind of information" (Thompson, 2018). Amid critical analyses of these "academy" or "library" metaphors vis-à-vis the nature of a platform's relationship with its users (Wyatt, 2021), it cannot be denied that with the platform's popularity – recently reported as the leading social media platform used by 97.2% of Internet users in the Philippines (We are Social, 2021) – and with the capacity to monetise content therein, it continually attracts content creators, now becoming a site holding millions of videos conveying people's everyday expressions, desires, and know-how.

The popularity of YouTube in the Philippines can be traced back to its pivotal role in stirring the career of aspiring celebrities in the Philippines. In 2007, Charice Pempengco, a young kid at that time, belted out a powerful Whitney Houston classic, "I Will always Love You," on a local reality talent show on television. While her performance landed her third, a fan uploaded a video of her performance on YouTube. The video became "viral" and generated millions of views from in and outside the Philippines. It is through the video that Charice rose to international stardom, initially invited to sing in a South Korean Talent show, *Star King* and eventually to appear as guest in prominent American talk shows, including *The Ellen DeGeneres Show* and *Oprah*. From there, Charice appeared in the popular American television series *Glee,* started performing across world-class venues, and worked with international stars in concerts and albums. Apart from Charice, other artists such as Arnel Pineda also landed an international career through YouTube. Arnel and his band named *The Zoo* once did a cover of classic Journey songs, including "Don't Stop Believin," and posted it on YouTube. The video led to Arnel's discovery to become the new lead singer of the legendary rock band Journey in 2007. Indeed, the success stories of Charice, who is more recently known as Jake Zyrus after coming out as a transgender male, and Arnel Pineda would not have been made possible without the networked connectivity and global exposure enabled by YouTube.

It is important to point out that the media has long been an important entry point for accessing fame and fortune, especially among celebrity-crazed Filipinos who navigate and negotiate the precarious and stringent social conditions in the Philippines. Ordinary Filipinos must deal with the everyday challenges of accessing social welfare benefits and job opportunities, with many ultimately deciding to move and work overseas (Rodriguez, 2010). Meanwhile, Philippine television has been perceived as a platform for earning profit and accessing stardom (A. Pertierra, 2018; J. Ong, 2015). This orientation of traversing precarious conditions has also been co-opted by Philippine entertainment media as reflected in capturing the Filipinos' aspirations to be discovered, overcome poor conditions, and help family members financially through a multitude of "showcase your talent" programs, game shows, and beauty pageants. Like television, YouTube, carrying an array of possibilities – global presence, celebrity status, and profit – attracts ordinary Filipinos to realise success via the creation of aspirational amateur videos. So along with the likes of Charice and Arnel who were "discovered" via YouTube, Filipino YouTubers have begun capitalising on the platform to visibilise their everyday and intimate lives or skills and knowledge, in the hopes of achieving a good and successful life amid the shortcomings of public institutions. As such, YouTube becomes a key site to both articulate those aspirations through content creation and for Filipinos to use YouTube as a vehicle to achieve those aspirations. In turn, these YouTubers, through their videos and the platform's features of facilitating communities of subscribers and viewers, continually cascade aspirations to others.

Following the successes of Charice and Arnel, YouTube tapped on the Philippine market by creating a local team in 2011. This opened opportunities for partnerships with existing media companies, such as GMA-7, ABS-CBN and TV5, as well as with local creatives like Filipino Society of Composers, Authors, Publishers (FILSCAP) and advertisers such as the multinational company Procter and Gamble (Olchondra, 2011). The local team also started the YouTube Partnership Program (YPP) in the Philippines, paving the way for Filipino content creators or YouTubers to monetise their content (Olchondra, 2011). The program has a sales team which deals with engaging content creators, promoting the use of analytics in their channels, managing targeted advertisements, and implementing copyright restrictions on content. At the time of the launch, a press release produced by a television network highlighted:

> Whether you want to be the next Charice, earn ad sales from original videos with the potential to "go viral," or promote a brand, the localised version of online video-sharing site YouTube will have you seeing numbers in a good way. (Olchondra, 2011)

In this statement, the allure of fame and possibly fortune through virality is implied. Particularly, it emphasises what networked connectivity can deliver, especially among YouTubers who consider the platform not only as a tool for cultural expression (Burges & Green, 2018) but also for monetisation (García-Rapp, 2017). It is also important to note that Google promoted the advantages of networked and global connectivity as "YouTube Philippines also gives local creators and brands the opportunity to increase exposure not only nationwide but globally as well" (Olchondra, 2011). Nevertheless, YouTube is framed as a space that affords profitability on a national and global scale.

This book presents a critical investigation of YouTube as an important site for examining Philippine modernity amid economic globalisation, although the ideas here may resonate with the experiences in other countries in the Global South and in facilitating reflections on digital cultures elsewhere. YouTube, a video and networked-based platform founded in 2005 in the United States and eventually bought by Google in 2006, has essentially become an integral part of the everyday lives of Filipinos. As one of the most highly subscribed social media platforms in the Philippines (We are Social, 2020), its contributions in shaping the Filipino's consciousness and aspiration necessitate a critical examination especially at a time of expansive digitalisation and global capitalism. Considered as one of the top markets for YouTube, the number of Filipino subscribers over the past three years increased by 20 times, while the number of channels with more than one million subscribers has increased 10 times from 2016 to 2019 (Mercurio, 2019). In the Philippines, which is once hailed as the social media capital of the world (Mateo, 2018), YouTube is also one of the major platforms that Filipino content creators utilise in creating online presence, branding, fostering communities, and sustaining social networks. After all, it is where the audiences are as YouTube becomes a one-stop space for Filipinos to seek information, follow a certain activity, or learn or improve a skill. Ultimately, YouTube is therefore set to be more deeply embedded in the fabric of Philippine society.

As the first book presenting a systematic and theoretical analysis of YouTube and digital cultures in the Philippines, we approach content and practices on the platform as shaped by socio-cultural, socio-economic, socio-political and socio-historical conditions. In a similar way, we locate YouTube's role in the transformation of the Philippine social and digital landscape, by paying attention to the often creative, playful, affective, and personalised contents produced by Filipino YouTubers as case studies of the brokerage dynamics of Philippine modernity in digital times. YouTubers capitalise on producing their own media content in a rich and convergent media environment (Jenkins, 2006). They are considered ordinary individuals who utilise media channels in visibilising their most intimate and personal lives in and through the media, which Turner (2009) conceptualises as the "demotic turn." Those who amass a significant viewership are considered "micro-celebrities" (Senft, 2018) who use digital communication technologies to enable interactions as well as generate and sustain popularity. Abidin (2015) also identifies them as "influencers," typically maximising the use of social media channels to generate thousands and millions of followers and generate profit. The performances visibilised through diverse content and even offline engagements by such influencers are anchored on certain branding strategies (Marwick, 2013) such as developing intimate (Abidin, 2015), raw, and affective narratives (Berryman & Kavka, 2017) typically shaped by entrepreneurial influences and guidelines (Banet-Weiser & Juhasz, 2011; Duffy, 2017). For this book, we focus on the dynamics of aspirational content creation and engagement practices of Filipino YouTubers, which remains understudied amid the growing studies examining the intersections of digital media, influencer culture, creative cultural production, and platformisation elsewhere.

We also approach YouTube content as part of the broader platform ecosystem, configured by its affordances and governance structures. The content produced and consumed online are shaped by platform logics, operating through the platform's business model and governance mechanisms (van Dijck, 2013). For instance, what appears in the interface of a platform is often a by-product of the "popularity principle." This means that the most shared, the most liked, and the most commented posts have a greater chance of appearing and being recommended on the platform. The visibility and access to certain content are then shaped by these recommendations. Reviglio and Agosti (2020) point out, for example, that YouTube's recommendation systems "drive more than 70% of the time spent in the video sharing platform and 90% of the "related content" is indeed personalised (p. 10). Tied to capturing online interactions, this modulates our attention on particular content over others (Bucher, 2012). Through such processes, connectedness is moulded by the platform's logics, including the interface and tags to record and present data, the algorithms processing the data, as well as the flows of profit through targeted advertising (Andrejevic, 2007; van Dijck, 2013).

Yet YouTubers are not only governed by these structures – they anchor their strategies on these structures to make their content more visible, to expand their communities, or to advance their personal and professional agenda. Thus, as we will show throughout the book, YouTubers' work lies in this interplay of YouTube's affordances and governance mechanisms with the content creators' strategic use of the platform that privileges them with visibility and renders relevance to their affective narratives that together curate Filipinos' YouTube-mediated socio-technical world.

By closely examining Filipino YouTubers' creative, diverse, and affective contents and how these contents are strategically promoted and circulated, this book examines how YouTube is mediating Philippine modernity in the digital era. As such, this book raises several questions that problematise how YouTube becomes an important node in the digital and social life of Filipinos. We ask, what representations by YouTubers on different aspects of Philippine society are presented and curated on this digital space? How do cultural, historical, economic, and political influences inform the ways of interrogating representations and practices on YouTube? How do technological affordances and platform logics shape the production and circulation of content on the platform? What do representations and digital processes on YouTube reveal about the positionality of the Philippines and its citizens in a globalising and networked economy?

To address these questions, we deploy the lens of "brokerage" (Stovel & Shaw, 2012), particularly extending this frame in the context of digital media and communications. YouTubers, as online content producers, have been previously framed as micro-celebrities (Senft, 2008), influencers (Abidin, 2015), and cultural intermediaries (Hutchinson, 2017), and now, we consider them as "brokers" of relatable persona, lifestyle imaginaries, relationships, and mobility tactics. Thus, we examine the content on YouTube and the micro-celebrity strategies engaged by YouTubers, not just as mediated performances of Filipino everyday life, but as conditions where diverse social transactions and aspirations of Filipinos are enacted, brokered, commodified, and negotiated.

In interrogating how online content creation is entangled with the platform's governance structures and broader economic, political, socio-historical, and technological conditions, we conceptualise YouTubers as brokers. Brokers similarly capitalise on micro-celebrity branding strategies to attract and engage audiences and generate profit, and use these and the platform's affordances to manufacture, sell, and "bridge" Filipinos to their economic, socio-cultural, and even political aspirations. As we will argue throughout the book, brokerage processes thrive within personally affective contents and strategic networks. Thus, the engagement of micro influencer strategies allows YouTube brokers to achieve a legitimate and relatable persona, put together disparate and useful relatable information as content, and facilitate affect and engagement crucial for the capturing of people's personal, economic, and political aspirations within the context of the platform's attention and business logics. This approach offers a critical focal point that widens our understanding of YouTube's growing role in the Philippines (specifically), but with potential relevance to other postcolonial societies (broadly).

Postcolonial Philippines in an Era of Neoliberal Globalisation

The Philippines provides a unique and interesting case study to critically examine the booming market of online content creation across the world. This lies in how socio-historical transformations have shaped the social, political, economic, and media terrain of the country. The Philippines is at the frontier of unprecedented changes influenced by the expansion of global markets, international relations and policies, as well as the advent of communication technologies. It is for these reasons that ordinary Filipinos have ventured into various opportunities, locally and internationally. However, these conditions are underpinned by the pre-existing hierarchy and divide in the nation-state, which favour and reinforce the privileged position of social, economic, and political elites. This, we argue, is crucial in analysing the conditions, practices, and outcomes of a networked, marketised and politicised environment of amateur online content producers.

As a start, we critically engage the Philippines' colonial past, which has been fundamental in setting the social, economic and political terrain that Filipinos navigate. Several Filipino scholars have argued that the current condition of the nation-state is deeply tied to a colonial history (Aguilar,

2014; Rafael, 2000; San Juan, 2011), highlighting that colonial legacy manifests in forms of domination, control, and marginalisation. For instance, racial and class hierarchies were reinforced during the Spanish regime. Colonisers and Filipinos with Spanish descent assumed privileged positions in the country, ascribing their whiteness to a social status, civility, and advancement (Rafael, 2000; Arnado, 2019). Indigenous Filipinos and dark-skinned individuals, meanwhile, were associated with barbarism (Rafael, 2000). This racial and skin colour discrimination have been deeply embedded in Philippine society, reflected in how Filipinos perceive standards of beauty, success, and upward mobility. Meanwhile, a different political, economic, and social life crystallised during the American rule, translating into new forms of domination and marginalisation. In contrast to the "divide and conquer" strategy of the Spaniards, the Americans deployed the tactic of "benevolent assimilation." In 1898, President William McKinley promoted paternalistic, family-oriented and affective relations between the U.S. and the Philippines (Aguilar, 2014). Filipinos were brought into the American education system, where they learned and practiced the English language and exposed to American culture and traditions. Scholars have pointed out that this immersion in the American education system has cultivated a neo-colonial consciousness in Filipinos, a privileging of the English language in education and professional communication and its recognition as the language of social elites, consumption of popular American media, and the performance of American traditions and cultures even after the American colonial period (Aguilar, 2014; Rafael, 2000). Notably, the deployment of the American education system in the Philippines also allowed for the exportation of trained Filipino nurses to the U.S. during and after World War II (WW2) (Aguilar, 2014). The U.S.-Philippine relations continued even after Philippine Independence in 1945 through sustained partnerships with U.S. officials, multinational companies, and transnational networks. These political and economic ties between the former colonisers and Philippine officials led to enduring hegemonic and colonial rule (Tadiar, 2004). For instance, U.S. troops remained in military bases in the Philippines even after the declaration of its independence, which sustained the prostitution of Filipino women and children in those communities (Tadiar, 2004). Ultimately, the positionality of colonial masters in the pedestal has been reinforced during the American period, most especially through benevolent assimilation.

By the 1960s, the Philippines fell into an economic crisis in the aftermath of WW2. Poverty, unemployment and underemployment worsened. Labelled as the "sick man of Asia," economic managers proposed the adoption of structural adjustment policies. Former President Ferdinand Marcos turned to the International Monetary Fund (IMF) and World Bank (WB) for loans (Aguilar, 2014). The Philippine government signed into international neoliberal policies such as the Washington Consensus, granting the U.S. a direct and indirect control of the Philippine economy (San Juan, 2009). This has led to the privatisation, deregulation, and liberalisation of various institutions and social welfare services, further undermining the unstable economic state of the country.

When large-scale labour emigration began in the 1970s, the economic conditions were worsened by the 1973 oil crisis (Asis, 2017). The economy could not keep pace with population growth, and the country needed to provide jobs and decent wages. To address the economic crisis, the Philippine government resorted to the further exportation of cheap and skilled labour through the Labour Export Policy (LEP) signed and issued under the Presidential Decree 422 in 1974 (San Juan, 2009). The labour code institutionalised overseas employment as a temporary-turned-permanent stopgap measure in addressing the country's economic problems (De Guzman, 2003). This paved the way for the creation of government institutions that oversee and manage the training and deployment of overseas Filipino workers, including the Philippine Overseas Employment Agency (POEA) and Overseas Workers Welfare Administration (OWWA). An overseas employment is aspired to by many Filipinos who lack access to social welfare services and work opportunities in the Philippines. According to the data produced by the POEA between 2015 and 2016, approximately 2.5 million Filipinos worked abroad, including 2 million land-based workers and 501 thousand sea-based workers (POEA, 2019). Although labour migration helps in supporting the accumulation of capitals needed by many Filipinos to escape poverty or support the education of left behind family members, it has many social costs, including physical abuse and inhumane living and working conditions for workers, as well as the disintegration of families (Rodriguez, 2010), Yet, despite its social costs, a significant portion of the local economy is driven by migrant remittances, which in turn also substantiates the economic benefits of labour exportation.

The migration trend has also activated the brokerage role of recruitment companies and informal actors that tapped diverse social networks and platforms to reach Filipinos, mostly women, who are aspiring for marriage migration. Along with the perception and aspiration that marrying a foreign partner can be a ticket to a more comfortable life, the growing statistics on marriage migration has caused anxieties about the welfare of Filipinos who marry foreign nationals through these transactions. A new law, RA10906 was enacted to strengthen the *Anti-Mail Order Bride Act of 1990*, and which considers the trafficking of Filipinos as brides or husbands through mail, in person, or over the Internet for the purpose of marriage or common law partnership. The problem with these state policies that are intended to address the welfare of Filipino migrants, however, is that local implementation mechanisms were often not in sync with national frameworks (Asis, 2017).

The Philippines reflects how a nation-state operates as a "broker" of cheap, surplus and exploited labour, engaging in neoliberal tactics on its conquest to address the nation's economic crisis (Rodriguez, 2010). Institutionalising this brokerage function is a web of social actors, offices, and institutions that facilitate the manufacturing of Filipino labour across the world (Guevarra, 2010; Rodriguez, 2010). Moreover, brokering is implemented by an oligarchic government, often adhering to colonialist and imperialist policies and generating more profit for those in power (Tadiar, 2004). Tadiar (2004) articulates the visions of the Philippine government for ordinary Filipinos as a "fantasy production" or an imagination that "is an intrinsic, constitutive part of the political economy" (p. 4). She writes:

> Fantasy production views the forms and dynamics of subjectivity produced and operating through contemporary international politics and economics, as emerging precisely out of dominant cultures of imperialism. Besides the orientalism in economics that persists in the world project of "development," logics of patriarchy, sexism, homophobia, and racism deeply inform and are generated by the practice of accumulation and power of postcolonial nation-states according to the tacit rules of the world system (p. 12).

Over the past years, the Philippine government has continued to broker Filipino workers to foreign markets (Guevarra, 2010; Rodriguez, 2010). The seamless exportation of human labour is well-established by the connections, control, and hierarchies wielded through policies during and after colonisation. Government and non-government institutions facilitate various training programs, serving as spaces for constant policing, monitoring and conformity to the ideal worker trope (Rodriguez, 2010). Overseas workers are encouraged to embody a certain subjectivity -- a supportive family member, a good citizen, or a competitive and entrepreneurial individual. These characteristics of an "ideal" overseas Filipino worker reflect a "neoliberal ethos" or sacrificing one's needs to provide for the family and the nation (Guevarra, 2010). This subjectivity is also promoted using mobile devices (Cabalquinto, 2014; Madianou & Miller, 2012) to fulfil duties and responsibilities to family through sending remittances and consumer goods (Cabalquinto & Wood-Bradley, 2020; Guevarra, 2010; Rodriguez, 2010). Nonetheless, construction of Filipino subjectivity, mobility, and labour is shaped by neoliberal and neo-colonial influences.

The Philippines is also embracing neoliberal globalisation, which is characterised by keeping up with other nation-states' expanding markets and strategies for profit accumulation (A. Ong, 2006). One of its major structural adjustments is partaking in the knowledge economy. By definition, the knowledge economy is the expansion of production processes that facilitate the creation of information and ideas (Radhakrishnan, 2011). In the Philippines, the knowledge economy manifests in the rise of business processing outsourcing (BPO) companies that delegate management, coordination, and processing schemes to individuals. Significantly, the boom of the call centre and medical transcription industry has manifested a seismic turn in how the Philippine government addresses its surplus labour (David, 2016; Fabros, 2018). Based on available statistics, over half a million Filipinos were involved in the BPO industry in 2017 (Philippine Statistics Authority, 2018), and the sector is a major contributor to the economy, projected to generate $29B in revenues in 2022 (IBPAP, 2020). Through BPO work, Filipinos do not leave the country but are compelled to imagine a foreign territory where they service their clients. Padios (2018) further notes that the competitive advantage of Filipinos in BPO work lies in their colonial roots, articulating the concept of Filipino/American relatability or the way the Philippines maintains its affinities and connections to the U.S. This concept demonstrates how Filipinos develop social capital and value through colonial legacies, which can also be negotiated when issues of racism and abuse emerge (Padios, 2018).

Like migrant workers, Filipinos in the BPO industry are described as "modern heroes" serving foreign clients, reducing unemployment, and bringing home much needed economic push, although this time not having to leave the country. Yet, although BPO work offers viable employment to many Filipinos such as acceptable pay, security of tenure, as well other benefits than their counterparts in the manufacturing and agricultural sectors, research has shown the physical and emotional burden experienced by local BPO workers. These included precarious working conditions characterised by long working hours, mental and emotional stress given the nature of work that required them to attend to irate customers on a daily basis, having to deal with constant night shifts, and mandatory overtime and holiday work (Fabros, 2016), among others.

More recently, the entanglement of technological innovations, further outsourcing of workers and services, and the widespread uptake of relatively affordable digital communication technologies have contributed to other technologically mediated service work in the Philippines. This outcome has also been aligned with institutional support on improving the country's digital landscape. This includes the development of the national broadband policy and promotion of ICT-related jobs through its *digitaljobsPH* program.

The Philippine government's promotion of digital labour as a viable source of income becomes the penultimate and current neoliberal strategy to address its surplus labour and a growing informal economy. This happened alongside the trend towards flexibilisation of work arrangements brought upon market transformations in the Global North (Soriano & Cabañes, 2020a), and which led to planetary labour markets (Graham & Anwar, 2019) involving labour platforms that match workers and clients across the globe. Online platform labour proved to be a favoured alternative for many workers who started to experience discontent from the challenging conditions attached to BPO work as well as the worsening traffic conditions in the metropolis (Soriano & Cabañes, 2020a). Now the label of OFW—formerly pertaining to overseas Filipino workers, is attributed to digital workers too—as Online Freelance Workers 2.0. As "OFW 2.0," Filipinos explore cloudwork platforms such as Upwork or Onlinejobs.ph that match them with overseas clients for dollar earnings. As we will show in the succeeding chapters, platform workers and emerging influencers join the government in the avid promotion and normalisation of platform labour as a viable employment alternative across social media. However, online freelance workers are not detached from difficulties and challenges, such as the constraints stirred by non-recognition of legibility of online work, overwork, isolation, limited potential for career advancement, lack of bargaining capacity, and other challenges connected to navigating the ambiguous platform environment (Soriano & Cabañes, 2020a, 2020b). It is through this point that Soriano and Cabañes (2020a) raise the issues surrounding how the Philippine government brands online freelance workers as "world-class workers" while eliding the deep-seated structural inequalities in the national labour economy.

Anchored with considering digital technologies as a site for economic gains, Filipinos' capacity to capitalise on available digital opportunities beyond the state's brokerage role has expanded. Aside from being platform workers, Filipinos are also actively creating monetisable content and market-ing products as entrepreneurs on social media platforms such as YouTube, while reviving old tensions of opportunity and precarity marked by past labour arrangements. The labour facilitated by these "digital opportuni-ties" also require new abilities and predispositions that its users need to accumulate and convert capital, even as sometimes this paradoxically reinstates inequalities and anxieties (Duffy, 2017; Marwick, 2018).

By mapping the transformations in Philippine history, we highlight the interplay of culture, expanding markets, technological advancements, and job insecurity and how it informs the process of labour, subjectivity, mobility, and marginalisation. Crucial in the entanglement of these elements is the process of how individual lives are tied to the country's colonial past that mediates the brokering of aspirations, mobility, labour, and even anxieties in modernisation, and which we discuss further in Chapter 2. In the next section, we discuss how the development of digital and mobile media in the Philippines builds the historical, cultural and social foundations of brokerage that manifest in online and social spaces like YouTube.

Digital Philippines

The explosion of digital technology and access opportunities ushered by some key global and local developments set the landscape for key transformations in digital communication in the Philippines. As Filipinos achieved greater geographical and social mobility along with the modernisation of Asian economies (Qiu, 2007), digital communication became more important than ever to cater to the sharp rise in informational, economic, social, and political demands of citizens. For instance, many Manila-based dwellers, for example, are local migrants from across the archipelago who moved to the city for employment aspirations (UNESCO, n.d.). Notably, the outward mobility from the regions and even the country for personal, familial and professional reasons enables separation but also creates imperatives for heightened digital connectivity. This condition is then addressed using mobile devices for sustaining relationships in a national (R. Pertierra et al., 2002) and transnational context (Madianou and Miller, 2012). Further, mobile phones also facilitate the enactment of identity, intimacy, and even political participation (R. Pertierra et al., 2002). In a non-proximate arrangement among Filipinos and their loved ones and social networks, digital devices facilitate the flows of goods, money and information, which mediate relationships, feelings, and aspirations. It is within this context of mobility, structure of social stratification, and a growing need to maintain personal, familial and even political connections that mobile and social media became popular.

Although the market for telecommunications in the country remains controlled by a few powerholders and connectivity one of the slowest in the world (Ookla, 2020), the high uptake is connected to telecommunication companies' rapid marketing of its products and services to tap the mobile Filipino market base with schemes attuned to the differing consumption capacities of Filipinos. There are two major telecommunication firms in the Philippines, including Globe Telecom and Philippine Long-Distance Telephone (PLDT). The latter owns two mobile communication companies, Sun Cellular and Smart Communications (previously known as Red Mobile). Globe Telecom and PLDT have been in competition in the local telecommunications market, although this might change as a third telecommunication company, DITO Telecommunity, was just given franchise by the Philippine government (Camus, 2019). Both PLDT and Globe sell a wide range of mobile Internet plans to accommodate a fast and changing market. More recently, mobile Internet is driven by the marketability of local brands that sell lower priced Android phones ranging from USD$25 to USD$300, such as *MyPhone, Cherrymobile, Starmobile,* and so forth. Post-paid telecommunications subscription offerings, along with the growing range of dealers, concept stores and kiosks spread across shopping malls and other public spaces that sell mobile devices with customisable features and software drive the tremendous mobile Internet market.

Telecommunication companies have been continuously reworking its products and services to suit the Philippine market across a broad income range. Crucial here is the offering of prepaid subscription, which caters to Filipinos from the lower to middle income classes, who experience the challenge of producing the requirements needed to open post-paid accounts. Prepaid sims with 4G or LTE Internet connectivity can be conveniently purchased for as little as USD$0.80 from the many convenience shops and variety stores spread across the country, and these sims are sometimes given away during company events or concerts. According to the International Telecommunications Union (ITU), there were 167.32 million mobile-cellular telephone subscriptions in the Philippines in 2019 (ITU, 2020), effectively more than the population and which implies multiple subscriptions for some sectors of the population. Another popular feature is the introduction of features allowing the purchase of mobile credit through *tingi* [purchasing in small increments], as well as autoload/e-load (over-the-air purchase of credit) and *pasaload*/share-a-load (over-the-air sharing of credit) which were made readily available to poorer segments of the Philippine society through micro-entrepreneurs (Soriano, 2019). To date, the *tingi* approach has also been deployed in the use of the Internet, social media and mobile applications. For instance, both Globe and Smart Telecoms have prepaid and postpaid products that bundle these services, tiered in terms of its combinations of texting, phone calls, and Internet packages (unlimited or capped to an amount or period), within and outside the networks, at home or mobile. Ultimately, these packages allow Filipinos to obtain "good enough access" (Uy-Tioco, 2019) to a range of online channels at home or on the go, depending on their capacity to pay. Smart and Globe, through their subsidiaries *Touch Mobile* and *Talk and Text* that target the lower-income market, now embed Facebook, YouTube, Instagram, Viber, TikTok, and other apps in their promotions for free data and low-cost or "unlimited calls, text, and data bundles." In some cases, telecommunication companies and microentrepreneurs rolled out *pisonet* units (Soriano, Hjorth, & Davies, 2019) in urban and rural (often low-income) neighbourhoods. These are coin-operated public access points akin to a *videoke* or arcade machine that allow for accessing calculated Internet time (i.e. 4 minutes) for a peso for those who cannot own personal computers or mobile devices.

This proliferation of creative plans and marketing schemes aligned with the boom in social media platforms, leading Filipinos to engage in various online activities, such as watching online videos, listening to music streaming services, podcasts, online radio, and so forth leading to being awarded with different titles. The Philippines was earlier regarded as the "texting capital of the world" (R. Pertierra et al., 2002), as the "social networking capital of the world" in 2011 (Stockdale & McIntyre, 2011), as well as the "selfie capital of the world" in 2014 (Wilson, 2014) where citizens spend the highest number of hours on social media (at 6 hours and 43 minutes on average) everyday (We are Social, 2020).

The rise of digital innovations is spurring on new business models and applications that find a wide range of appropriations in a developing economy with a high level of communication skills and, ironically, social stratification (Lorenzana & Soriano, 2021). Ultimately, smart phones have become tools to sustain familial or friendly relations (R. Pertierra et al., 2002) as well as intimate relationships through dating apps (Cabañes & Collantes, 2020; Labour, 2020). It has also been used to sustain cultural identity and religious affinities (Ellwood-Clayton, 2005). Moreover, social media channels have been utilised to navigate and cope with the impacts of disasters (Crisostomo, 2020; David, Ong, & Legara, 2016; McKay & Perez, 2019). Globally, digital communication technologies have connected dispersed family members (Cabalquinto, 2018a, 2022; Madianou & Miller, 2012), offering ways of sustaining mothering roles, transnational fathering, as well as the fulfilment of filial obligations (Cabalquinto, 2018a, 2022; Cabañes, J. & Acedera, K., 2012; Parreñas, (2001); Uy-Tioco & Cabalquinto, 2020). Notably, the rise in the use of social media for political engagement and communication is also growing, leading to the insertion of the struggles of minority and marginalised peoples in national narratives, while in turn also fuelling the populism that has gained international attention.

The continued growth of interactive applications reconfigures the Filipino consumer base as active and dynamic publics choosing from a range of communicative options and relational possibilities (Madianou & Miller, 2012) and as creators of content, seeking opportunities for heightened self-expression, recognition (Lorenzana, 2016), and attention (Cabalquinto & Soriano, 2020). Yet, these pose new questions and debates. One such issue that has emerged is the impact of digital technologies on cultural forms and expressions of identity, community, and nationhood (McKay, 2011; Crisostomo, 2020). Along with openings for creative and artistic expressions is also the rise of disinformation and populism (J. Ong & Cabanes, 2018; Curato, 2018), incivility and scandals (Lorenzana, 2021), cyberbullying, cancel culture, and influencer economies (Cabalquinto & Soriano, 2020) that have taken the country into uncharted political and social terrain. The opportunity to insert new voices into public conversations facilitated by spaces such as YouTube coincide with prejudice, bias, or hate. Further, communicative relations on social media are enacted as social, economic, and political transactions that are monetisable, feeding the capitalist ethos of capital accumulation, influence, and reputation (Athique, 2019). With the explosion of digital technology and access opportunities ushered by key global and local developments described above, digital communication in the country is heightened as it is commercialised through the conversion of every post and engagement into data and data as commodity form.

As discussed earlier, the Philippines' social and economic conditions are entangled with political structures and historical influences. Developments in digital communication and the conditions and structures align with colonialist dreamworks and fantasies. The next section elucidates how political, economic, and socio-historical factors shape the contours of the fast-evolving media ecosystem in the Philippines. This is of great importance especially in understanding and examining the practices, performativity, and politics of content creation as these evolve within a broader media ecosystem.

On/Off Air? From Mainstream to Online Media

Despite a growing number of studies on micro-celebrity in Asia (Abidin, 2015, 2016; Abidin & Brown, 2019) and in the Western world (Duffy, 2017; Marwick, 2013; Poell, Neiborg & Duffy, 2022; Senft, 2008), little study has been made on the flourishing presence of online content creators in the Philippines (Shtern, Hill, & Chan, 2019). Our research contextualises the representational politics of content produced by YouTubers in the Philippines, highlighting how texts and practices are shaped by and deeply tied to the Philippine's cultural, economic, political, and socio-historical domains. To understand the politics behind texts and mediated practices, we first map the media landscape in the Philippines. In particular, the high penetration of traditional media and particularly television in Philippine households, as well as the strong cultural influence of television programs in articulating values and aspirations among Filipinos, require examining how its functions are extended, sometimes replicated, and further expanded by content creators on YouTube.

The analysis of content creation in a visual medium such as YouTube necessitates a complementary understanding of the popularity of television in the Philippines. Philippine television, with its accessibility and cost friendliness, has been an iconic medium in Filipino households. It circulates contents that mediate Filipino values and aspirations. Yet, it also reflects the existing social differences and hierarchy in Philippine society. It is often populated and governed by the middle-class and elite Filipinos and mestiso/ mestisas. As a business, Philippine television negotiates the social hierarchy by producing programs and contents that align with the interest, taste and even aspirations of the target audience, typically the *masa* or individuals belonging to lower income classes. Here, televisual spectacle is transformed into the blending of the upper and lower-class Filipinos in television formats (A. Pertierra, 2018), such as game shows, reality TV, and so forth. In examining the political and economic dimension of media and visual texts in the Philippines, scholars have deployed the lens of patron-client relations (J. Ong, 2015; A. Pertierra, 2018, 2021). This means that gestures, practices, and speeches are delivered by social actors in exchange for a wide range of capitals – social, economic, and political. A key work in Philippine media studies that articulates patron-client relations is by Anna Pertierra (2018). Her ethnographic study of a long-running noontime show, *Eat Bulaga*, unpacked how celebrities, who belong to the middle and upper class, can be one with the masses through performative acts and a range of gimmicks. Philippine television's dominant audience base belongs to lower-middle classes and the urban and rural poor. A. Pertierra (2018) argues that the masses, who were typically the market and the contestants of the show, express their experiences of poverty through stories of suffering as well as humour and mimicry embedded in the shows. This enables them to generate a sense of agency in the form of visibility as well as economic gains by bringing home instant cash prizes. This negotiation of social differences is echoed in the ethnographic study of Jonathan Ong (2015) on how Philippine news and several game shows represent the suffering of poor Filipinos. Ong underscored how the everyday sorrows and suffering of ordinary Filipinos are common themes to attract audiences' attention and further engage advertisers and business enterprises. By interviewing producers and journalists, J. Ong (2015) unpacked the tensions between the intersections of economics and morality in broadcast journalism, highlighting how the spectacle of suffering is often utilised to articulate the deservingness of aid and even airtime.

The practices of Philippine television essentially show how emotionally charged stories have become valuable, palatable and profitable contents. But more importantly, both studies of A. Pertierra (2018) and J. Ong (2015) also highlighted the ways through which various capitals are mobilised in Philippine television. More specifically, the visibility of television hosts in helping the masses on national television becomes potent mechanisms for admiration among masses, which can eventually translate to popularity or campaign support and electoral votes for celebrities who run for public office (A. Pertierra, 2017). For the former, visibility can lead to multi-million contracts, product endorsements, international tours, and so forth. For the latter, several Filipino celebrities, such as Tito Sotto, Bong Revilla, Noli De Castro, Vilma Santos, Joseph Estrada, and many more, have eventually occupied key positions in government, including the Presidency, Vice Presidency, Senate, and Congress. Ultimately, entertainment media stirs political capital (R. Pertierra, 2020; A. Pertierra, 2020), which is enabled by patron-client relations or how connections and affiliations to certain personalities can offer access to upward social mobility.

A patron-client relation also reinforces power structures in Philippine television. The experience of poverty is marked by representations in the media and the consumption of a range of media contents (A. Pertierra, 2018; J. Ong, 2015), which are appropriated for monetisation. In Philippine entertainment media, celebrities and their networks direct participants to act, dance, sing, laugh or display their uncanny talents while offering solace through instant gifts or cash prizes, underlining the marginalised status of the guests and the role of the media networks in facilitating empowerment. This same trope – that economic hardship can be alleviated through opportunities provided by the media and in particular through "getting discovered" – is a recurring feature of teledramas and movies too, and as we will show in this book, resonate in many YouTube videos as well. Yet, entertainment and visibility does not necessarily undermine or change systems that produce inequalities. As A. Pertierra (2018) emphasises: "But this increased access to material abundance in no way overcomes – and in some cases rather continues to emphasise the marginal or subaltern status of these groups" (p. 2). Ultimately, the affective performances and the consequent access to capital facilitated by Philippine television indicates that the media is challenging but also perpetuating hierarchical structures. As to be presented in this book, we showcase how an online video-based platform such as YouTube becomes an extension of these negotiations of social differences in the Philippines.

The middle class, elites, and politicians take the media stage, taking advantage of the active presence of ordinary Filipinos to advance political agendas and commercial gains. These individuals are privileged in their position because of their access to diverse capitals and their strategies for maintaining their position. So with the boom of digital communication technologies, Filipinos are ushered into a new space that offers possibilities yet also reinforces domination and marginalisation. The online space serves as a melting pot for crafting aspirations, anxieties, frustrations, and accessing resources that remain inaccessible in Philippine society. However, to date, limited studies have been conducted to examine the emerging media practices of ordinary Filipinos utilising online media channels to broadcast their lives (Shtern et al., 2019). It is through this gap that our book intervenes.

In the first instance, we build on existing studies that have highlighted how patron-client relations sustain the "harmonious" dynamics of different and class-based individuals (J. Ong, 2015; A. Pertierra, 2018). In online spaces, Filipinos who use online channels typically benefit from a patron-client engagement, as reflected in the form of exchanging recommendations, advice, and networking. They capitalise on affective and personalised storytelling to sustain social connections among their networks as well as translate interactions to platform-based profits in the forms of subscription, likes, views and shares. What the current media and digital environment in the Philippines shows is that ordinary Filipinos have been given access to broadcasting tools such as YouTube to curate, produce, circulate, and monetise their own content. These individuals contribute to what A. Pertierra (2021) calls, "entertainment publics" or "comedic, melodramatic and celebrity-led content" that generate "networks of followers, users and viewers whose loyalty produces various forms of capital, including in notable cases political capital" (p. 2). Through the mainstreaming of relatively affordable smartphones, accessibility of mobile social media and applications, as well as filming tools and software, an ordinary Filipino – as an influencer – can partake in practices of visibility and curation, while weaving aspirations and capitalising on monetisation possibilities afforded by the platform. The brokering then happens when their stories and affective performativity are produced, circulated, and consumed in a networked environment. Yet, the contradiction of occupying an online space is the reinforcement of market logics and an internalised neo-colonial mentality. For this study, we refer to the role of a video-based channel, YouTube, to expose the paradoxical conditions enabled by the digitalisation of Filipinos' lives in a neoliberal and postcolonial context.

YouTubing the Philippines: The Analytic Lens of Brokerage

This book focuses on YouTube as a locus for producing and circulating distinct creative, personalised, networked, commodified, and postcolonial contents. In this section, we highlight why it is important to study You-Tube in the context of the Philippines, and what significance this offers in understanding how individuals navigate the impacts of neoliberal and colonialist globalisation in the digital age. Further, our study offers the lens of brokerage as a critical perspective for examining the paradoxes surrounding representational practices in the neoliberal and marketised space of YouTube. The key aspects of "digital brokering" will be discussed in Chapter 2.

From its inception, YouTube has promoted and capitalised on encouraging users to engage in authentic self-expression and community building. Originally, it was introduced as an online tool that allows "ordinary individuals" (Strangelove, 2010) to broadcast their lives and demonstrate ways of cultural and personal expressions (Burgess & Green, 2018). At the time when YouTube was launched, popular genres of content involved banal, personalised, and vernacular creativity (Burgess & Green, 2018). Here, Strangelove (2010) has highlighted that "extraordinary" videos made by ordinary people typically captured personal and domestic exchanges. Strangelove considered YouTubing as a form of domesticated media practice, historicising how family members utilise a camera to capture, curate and archive everyday familial content.

Now, we see the rapid growth of diverse, multicultural, amateur and professional content creators from around the world who harness these platforms to develop and promote their own brands, engage in content innovation, and cultivate communities and followers. YouTube has brought into the fore the diverse creative content of cultural interest that were previously relegated to the intimate or private domain, such as beauty, eating, cooking, mothering, and other aspects of domestic and intimate life often shot through amateur videos (Burgess & Green, 2009; Cunningham, 2012; Kumar, 2016). From its minority and alternative media roots (Jenkins, 2006), YouTube has given rise to a wide user-created content community, and also now considered to be an important site for amateurs, micro-celebrities, small entrepreneurs, and even large companies for pushing their content for wide exposure and with monetary benefits. In light of the relatively frictionless global reach of various forms of "social media entertainment" such as YouTube (Cunningham & Craig, 2017; Cunningham) and the growing genre of YouTube videos that convey multiple aspects of everyday life, it is important how even for many Filipinos, "especially young viewers, this is what television is, now" (Craig, & Silver, 2016, p. 71).

While YouTube may be experienced as television for many, the platform brings in affordances that present unique forms of engagement with its users. Beyond the affordances of content creation and sharing, it is the networked connectivity of YouTube that has also paved the way for generating a sense of connectivity among its users, which also works for facilitating the potency of its content (Burgess, 2011). YouTube's architecture and design interface facilitates social interactions and connection among content creators (Burgess & Green, 2009; Lange, 2014), as well as among subscribers, fans, and visitors (García-Rapp, 2017; Lange, 2014). Amateur video production establishes connections and relationships among individuals on YouTube (Lange, 2009), while spoofs, parody and viral videos further encourage participation among larger audiences (Burgess, 2008). The online space has enabled confessions and coming out videos (Alexander & Losh, 2010) as well as launching political commentaries and advocacy campaigns via video mashups (Edwards & Tryon, 2009). Moreover, video blogs have mobilised individual expression and harness a space for collective and cultural expressions (Chu, 2009). Prominently, beauty vloggers capitalise on deploying intimate narratives and tutorials to increase viewership and generate profit (Berryman & Kavka, 2017). YouTubers also exploit key features of YouTube, such as encouraging viewers to hit the subscribe button or click the notification bell to establish connections with subscribers and viewers. The YouTube live feature affords the elicitation of audience engagement in real time, right when stories or issues are at their peak. With YouTube live, YouTubers have the opportunity to become "live broadcasters," (Soriano & Gaw, 2021) where they can give "shoutouts" to acknowledge their viewers and read their comments and questions aloud – a feature common in Philippine TV and radio broadcasts. YouTube's affordances allow content creators to blend TV, radio, and social media, producing content that is not restricted by the limitations of bandwidth or airtime while maximising audience-engagement strategies to make these content more dynamic and affective. Through this culture of content creation, circulation and engagement, YouTube becomes the anchor for different forms of cultural and associational expression among diverse publics (Cunningham, 2012; Jenkins, Ford, & Green, 2013). Its multiple affordances offer many possibilities for users to cluster around video content and channels across different topics and genres, including racialised, gendered, and class-based narratives (Strangelove, 2010).

Significantly, YouTube's transformation has also been informed by the operations of business models and data governance. Through partnership programs, individual users are afforded with the ability to monetise popular content (Burgess, 2011; van Dijck, 2013). This program complemented how media corporations have begun utilising YouTube as an alternative channel for distributing content, which then generated criticisms from amateur video producers who consider YouTube as a democratised, open, and alternative space for cultural production (Burgess & Green, 2009). In some cases, YouTube is criticised for promoting authenticity, vernacular creativity, and community-building following market logics (Burgess, 2015; Burgess & Green, 2009). As Burgess and Green (2009) note:

> YouTube is, and has always been, a commercial enterprise. But it has always been a platform designed to enable cultural participation. Despite all the complexity of its professional media ecology, the inclusiveness and openness of the YT promise that "anyone" can participate is also fundamental to its distinctive commercial value proposition. This is what we mean when we say that, for YouTube, participatory culture is core business. (p. 123)

This book approaches YouTube both as a storehouse of cultural expressions and a platform.

It serves as a tool for forging and maintaining personal and social connections. It allows individual users to present and curate their everyday experiences, movements, and social interactions and enact multiple ways of being and aspiring. Online performativity and interactions reflect what Burgess (2007) refers to as "vernacular creativity" or practices that capture the intimate, mundane, creative, and playful activities of individuals. Notably, as YouTube affords monetisation, individual users have also begun using the platform to generate profit. In this vein, amateur content creators transition to become influencers. These individuals deploy a range of communicative strategies to develop and sustain a connection between oneself and the audience (Abidin, 2015; Duffy, 2017). Utilising performative and affective narratives (Berryman & Kavka, 2017) has become a key branding strategy (Duffy, 2017; Marwick, 2013), which also raises concerns on constantly juggling being authentic vis-à-vis crafting a certain persona that meets the demands of the audience or a corporate brand (Duffy, 2017; Marwick, 2013). These strategies also involve varieties of "affective labour" (Berryman & Kavka, 2017) that go into producing such displays. This is especially notable given the blending of private life experiences into the construction of videos and in the process of sustaining a community of followers on social media. Where YouTubers expose their own vulnerabilities and personal successes into their branding strategies, they also emotionally expose their own aspirations and emotions, crucial to achieve and sustain authenticity on the platform and "cement ties of intimacy" with their followers (Berryman & Kavka, 2017, p. 96). This includes the "relational labour" or the effort that goes into and beyond managing others' feelings and maintaining connections that can consequently boost one's earning potential, and which pertains to the "complementary dialectics of personal relationships and professional labour" playing out in the ever-changing flow of everyday interaction on social media (Baym, 2015).

To date, YouTube, operating its partnership programs on a global scale, is primarily considered a platform that capitalises on its interface technology, data governance, algorithms, and business models. This book then asks, what do we know about YouTube in the context of the Global South and the Philippines in particular? What types of content are produced and circulated in an online space? What do online processes and content reveal about the ways the Philippines navigates a digitally mediated and global economy? In responding to these questions, we propose brokerage as an analytical lens.

We anchor our conception of brokerage to the fast-evolving relations between systems, markets, institutions, and digitalisation. Brokerage is defined as "the process of connecting actors in systems of social, economic, or political relations in order to facilitate access to valued resources" (Stovel et al. 2011, p. 141). This conceptualisation complements the propositions surrounding the practices of micro-celebrities, influencers or cultural intermediaries, individuals who often act as middle person between a brand, private and government institutions, and audiences. However, our conception of brokering is expansive, situated, and reflexive, highlighting how representations and practices in digital spaces are deeply linked to consequences of historically-situated economic globalisation. As we have shown earlier, media practices in the Philippines are informed by hierarchical structures that are created and reinforced by the logics of markets, neoliberal ethos, and colonial influences, and which all work to sustain the operation of brokers. By identifying the diverse and often intertwined factors that shape the types of content and affective performances online, we locate how the ordinary Filipino is visibilised and positioned in a digital realm. As to be presented in this study, we focus on the affective content and digital strategies that reflect the mechanisms of brokerage as both, echoing the words of Duffy (2017), practice and ideology.

By engaging the lens of brokerage in the context of YouTube, we seek to illuminate the desires and aspirations enabled and bridged by content reference, cultures, and practices of Filipino YouTubers. More than exposing the diverse, personalised, and localised digital practices, this book takes into consideration the overarching historically situated political and economic forces that inform media representations and practices. In this vein, YouTube is not only approached as a platform. Rather, it is examined as an intermediary or broker of economic, political, and social goals and aspirational lifestyles through the ways YouTubers perform, broadcast and monetise everyday life in a postcolonial state. As a form of networked brokering, we show how YouTubing subverts and reinforces ideal subjectivities, intimate relations, labour practices, and political expressions within a neoliberal and postcolonial domain. As our case studies unfold in the succeeding chapters, we aim to provide a critical entry point in analysing the impact of digitalisation on the everyday life of individuals who remain neglected and even exploited by the nation-state in a globalising and networked economy.

Methodological Approach

Online videos are considered textual trails in an online space, and media texts can be analysed as cultural expressions with interpretive meanings shaped by diverse contexts. We build on earlier works employing discursive textual analysis of visual texts on social media (Banet-Weiser 2011, Dobson, 2015; Holmes, 2016; Jancsary et al., 2016, pp. 180–204; García-Rapp & Roca-Cuberes, 2017; Strangelove, 2010) to examine the "recurrent narrative and aesthetic structures" embedded in the videos, including a focus on "temporal organization, editing, image, sound" (Holmes, 2016, p. 7), as well as branding strategies (Abidin, 20015; Berryman & Kavka, 2017; Duffy, 2017) employed by the content creator. We examine this orchestration of visual elements and communicative strategies that constitute a particular social reality embedded in text while also situated in a specific cultural context and dynamics of the platform.

This book is based on collecting and analysing selected YouTube videos by amateur content creators promoting a range of aspirations – white skin, interracial relationship, world-class labour, and progressive political governance. YouTube hosts millions of videos of a broad range of themes. We chose these four topics for our case studies because we felt they represented a spectrum of aspirations brokered on YouTube ranging from beauty and self-esteem (skin whitening), intimacy and social mobility (interracial relationship), economic opportunity (online freelancing and digital labour), and nation-building and progress (politics and governance). These four themes also pertain to different subjectivities: a feminised identity, a romantic partner, a world-class worker, and a patronising citizen, that encompass the multiple facets of the Filipinos' emplacement in global modernity. We used Google Trends to identify commonly searched keywords for each theme [i.e. *pampaputi ng kilikili* (underarm whitening) for skin whitening, "LDR" for interracial relationship, "online freelancing" for digital labour, and so forth]. In identifying videos for analysis, we selected amateur videos, excluding videos that are produced by professional institutions or media organisations. Because we are capturing amateur content creators who have amassed significant influence on the platform, the search results were filtered in terms of views and channel subscribership (Altmaier et al., 2019), and specifically those with above 1,000 subscribers and 4,000 hours of accumulated watch time for the past 12 months, which is YouTube's requirement for monetising content on its platform. From the resulting amateur videos and channels emerging from our search with the highest views and subscriptions, we closely examined the videos' aspirational content, discursive styles, community and credibility building strategies, and platform engagement tactics. To identify the aspirational tropes and strategies, we used thematic analysis (Flick, 2011) to surface emerging themes that are then discursively analysed in terms of how these aspirations are underpinned by gendered, racialised, classed, and postcolonial realities. Conducting a critical analysis of the videos unravelled the tropes in relation to the brokerage of aspirations embedded in content production and engagement practices on YouTube (García-Rapp & Roca-Cuberes, 2017; Lange, 2009). However, in the process of our analysis, we attempted to get beyond the level of particular examples or themes, and to gain some perspective on YouTube as a mediated cultural system (Burgess & Green, 2018) and its role in the everyday lives of Filipinos. There are some methodological variations in each chapter. For example, thematic analysis was complemented by interviews with some influencers for Chapter 5. An expanded discussion of the methods is presented in each chapter. The project obtained ethics approval from De La Salle University (DLSU-FRP.013.2019-2020.T2.CLA).

Organisation of the Book

This book has seven chapters, including this one, which characterise the social, economic, political, and technological transformations surrounding the brokerage of aspirations in neoliberal and postcolonial Philippines on YouTube. In Chapter 2, we present the theoretical framework of the research project. It extends the concept of brokerage in the digital realm, embedding brokering practices within existing social, economic, political, and historical conditions. Rather than emphasising brokering only as a strategy deployed by the Philippine government in response to globalised capitalism (Rodriguez, 2010), we argue that brokering operates as a result of deep internalisation and constant negotiation of individuals who perceive YouTube as a fundamental source of investment, mobility, networks, and capital. We problematise brokering through the lens of postcolonialism, emphasising how being a global and tech-savvy Filipino citizen becomes a marker of potential, imagined, or actual upward and networked mobility among ordinary citizens. Although an online space like YouTube can imply a democratisation of communicative capacities, we also identify YouTube as a contested site, in which individuals negotiate conditions of social mobility and immobility, progress and precarity, and belonging and exclusion.

Chapter 3 explores the role of YouTube in enabling YouTubers to aspire and embody a white subjectivity. More specifically, it presents how Filipina YouTubers promote underarm whitening as an everyday, banal yet critical practice in enabling an ideal femininity and beauty. We approach the broadcast of underarm whitening as a form of a temporary "cultural whitening" or wanting to be white. As a form of colour consciousness, having a white or fair skin is shaped by Philippine colonial history. Here, whiteness is ascribed with privilege, social status, civility, and upward mobility. Of particular interest in this chapter is a discussion on how the visibility of Filipina women and their skin whitening practices signal a postfeminist subjectivity – enacting empowerment through digital, neoliberal and entrepreneurial practices. Ultimately, an examination of content on underarm whitening demonstrates the brokering of ideal standards of feminised and racialised subjectivity, the visibility of which generates imaginaries and aspirations for attractiveness and marketability in and beyond online spaces.

Chapter 4 unpacks the brokering of interracial and mediated intimacies on YouTube. We focus on the curated stories of Filipina YouTubers who met their intimate partners via online channels. Analysed as an extension of marriage migration brokerage enacted by formal or informal recruitment agents, we show how YouTube becomes a site for the performance of interracial intimacy while cascading imaginaries, aspirations, and importantly, know-how, of finding a white foreign partner and achieving a successful interracial and intimate relationship. As a form of embodying cultural whitening through marriage and eventually having a mixed-race child, intimate narratives broker the notions of an intimate, authentic, and happy relationship. Paradoxically, the stories of Filipina YouTubers serve as "countererotics" or the challenging of the sexualised representation of Filipino women in an interracial relationship. Yet, representational politics can remain especially when performativity reinforces gendered subjectivity – a woman who is idealised as caring and domesticated. Further, it is also this gendered performativity that is commodified in an online space. In a way, performing intimate and interracial relations indicate a postfeminist subjectivity, as reflected in investing in digital practices to commodify and broker gendered, racialised, and intimate encounters and aspirations.

Chapter 5 underscores how YouTube is engaged as a platform for mediating digital labour through "skill-selling." Through interviews with platform workers and analysis of "skill-making" content on YouTube, the chapter presents YouTube as a space where YouTubers can showcase their capabilities to obtain a captive market and attain celebrity status as "global knowledge workers." Using YouTube provides an opportunity for digital labour influencers and skill-makers to deliver training to aspiring platform workers who seek to earn dollars while working at home, while also crafting imaginaries and ideals of success in the platform economy. In effect, YouTubers, through the brokerage of skills and promotion of the viability of digital labour, perform the role of local matchmaker between aspiring workers and digital labour platforms. Situated within the frames of the gig economy, this chapter offers a critical insight on digital and flexible labour in the Global South and the role of YouTube in brokering labour relations and economic aspirations.

Chapter 6 examines how YouTube facilitates the brokering of a political agenda through historical revisionism. We analyse the content and strategies of YouTubers advancing revisionist narratives about the brutal history of Martial Law under former and authoritarian President Ferdinand Marcos, as a pathway for the cleansing of the Marcos legacy and in preparation for the campaign of Marcos' son, Ferdinand (Bongbong) Marcos, Jr,. as President of the Republic. Considered as political brokers, YouTubers take advantage of the platform's affordances and porous governance structures by creating content and a network that build, propagate, and cement their political narratives without being subjected to the same scrutiny of and by traditional gatekeepers. The crass online performativity of political brokers is utilised to broker national aspiration of progress and economic security patterned after the West. We argue that democracy can be threatened and undermined especially when a profit-driven and emotionally laden propaganda re-casts a turbulent political history by staging a mythic agentic space for online supporters. Through the lens of brokering, we can hypothesise how Philippine politics will be further reconfigured by new faces created and influenced by an ongoing production and circulation of unregulated disinformation content on YouTube.

Concluding this book is a discussion on some of the key contentions on brokering feminised subjectivity, interracial relations, world-class labour, and partisan politics. The closing chapter also leaves the readers with some future research directions, including ways of rethinking YouTubing as a form of platformisation of everyday life in the Global South. Significantly, at the centre of this chapter is a critical reflection on the contradictions of a digital life in the Philippines. It does not only reiterate the affordances and possibilities that are activated through engagement with social media (broadly) and YouTube (specifically). It also emphasises the ruptures and tensions that YouTubers have to constantly manage as reflected in their affective performativity, networked engagement, and strategic online content creation. We argue that navigating a digital terrain is understood as symptomatic of the inequalities produced through the broader systems of neoliberalism and colonial legacy in the Philippines. Within the frames of a global economy and further flexibilisation and informality of work, the individual becomes responsible for his or her own undertaking given the lack of available government-run social welfare programs and public services, and the challenges of navigating information amid the complexity of local and foreign bureaucracies. However, through YouTube, the individual can position oneself as an entrepreneurial and global persona or obtain imaginaries of possibility to thrive and achieve their aspirations. Paradoxically, engagement in an online space is co-opted for commercial interests and even partisan politics. Nonetheless, we propose a much-needed critical lens to investigate how a personalised, creative, and playful space for self-expression and community building may obscure the often-invisible injuries of economic globalisation.

References

Abidin, C. (2015). Communicative ❤ intimacies: Influencers and perceived interconnectedness. *Ada: A Journal of Gender, New Media, and Technology, 8*, 1-16. doi:10.7264/N3MW2FFG

Abidin, C. (2016). Visibility labour: Engaging with influencers' fashion brands and #OOTD advertorial campaigns on Instagram. *Media International Australia, 161*(1), 86–100. doi:10.1177/1329878X16665177

Abidin, C., & Brown, M. L. (2019). *Microcelebrity around the globe: Approaches to cultures of Internet fame* (First ed.). Emerald Publishing Limited.

Aguilar, F. J. (2014). *Migration revolution: Philippine nationhood and class relations in a globalized age*. Ateneo de Manila University Press.

Alexander, J., & Losh, E. (2010). A YouTube of one's own? 'Coming out' videos as rhetorical action. In M. Cooper & C. Pullen (Eds.), *LGBT identity and online new media* (pp. 37–50). Routledge.

Altmaier, N., Beraldo, D., Castaldo, M., Jurg, D., Romano, S., Renoldi, M. Smirnova, T., Seweryn, N., & Veivo, L. (2019). *YouTube tracking exposed: Apps and their practices*. Retrieved 6 July 2020, https://YouTube.tracking.exposed/trexit/

Andrejevic, M. (2007). Ubiquitous computing and the digital enclosure movement. *Media International Australia, Incorporating Culture and Policy* (125), 106–117.

Asis, M. (2017). The Philippines: Beyond labour migration, toward development and (possibly) return. Migration Policy Institute. Retrieved 1 March 2021, https://www.migrationpolicy.org/article/philippines-beyond-labour-migration-toward-development-and-possibly-return

Athique, A. (2019). *Integrated commodities in the digital economy. Media, Culture & Society, 2*(4), 554–570. doi:10.1177/0163443719861815

Banet-Weiser, S. (2011). Branding the post-feminist self: Girls' video production and YouTube. In M. Kearney (ed.) *Mediated girlhoods: New explorations of girls' media culture* (pp. 277–294). Peter Lang Publishers.

Banet-Weiser, S., & Juhasz, A. (2011). Feminist labour in media studies/communication: Is self-branding feminist practice? *International Journal of Communication (19328036), 5*, 1768–1775.

Baym, N. (2015). Connect with your audience! The relational labour of connection. *Communication Review, 18*(1), 14–22. doi:10.1080/10714421.2015.996401

Berryman, R., & Kavka, M. (2017). 'I guess a lot of people see me as a big sister or a friend': The role of intimacy in the celebrification of beauty vloggers. *Journal of Gender Studies, 26*(3), 307–320. doi:10.1080/09589236.2017.1288611

Bucher, T. (2012). Want to be on the top? Algorithmic power and the threat of invisibility on Facebook. *New Media & Society, 14*(7), 1164–1180. doi:10.1177/1461444812440159

Burgess, J. (2007). Vernacular creativity and new media (Doctoral dissertation), Queensland University of Technology, Brisbane, Australia. Retrieved 26 October 2020, http://eprints.qut.edu.au/16378/1/Jean_Burgess_Thesis.pdf

Burgess, J. (2008). 'All your chocolate rain are belong to us?' In G. Lovink & S. Niederer (Eds.), *Video Vortex reader responses to YouTube* (pp. 101–110). Institute of Network Cultures.

Burgess, J. (2011). User-created content and everyday cultural practice: Lessons from YouTube. In J. Bennett & N. Strange (Eds.), *Television as digital media* (pp. 311–331). Duke University Press.

Burgess, J. (2015). From 'broadcast yourself' to 'follow your interests': Making over social media. *International Journal of Cultural Studies, 18*(3), 281–285. doi:10.1177/1367877913513684

Burgess, J., & Green, J. (2009). *YouTube: Online video and participatory culture*. Polity Press.

Burgess, J., & Green, J. (2018). *YouTube* (Second ed.). Polity Press.

Cabalquinto, E. C. (2014). At home elsewhere: The transnational kapamilya imaginary in selected ABS-CBN station IDs. *Plaridel: A Journal of Philippine Communication, Media and Society, 11*(1), 1–26.

Cabalquinto, E. C. (2018a). 'I have always thought of my family first': An analysis of transnational caregiving among Filipino migrant adult children in Melbourne, Australia. *International Journal of Communication, 12*, 4011–4029.

Cabalquinto, E. C. (2018b). 'We're not only here but we're there in spirit': Asymmetrical mobile intimacy and the transnational Filipino family. *Mobile Media & Communication, 6*(2), 1–16.

Cabalquinto, E. C. (2022). *(Im)mobile homes: Family life at a distance in the age of mobile media.* Oxford University Press.

Cabalquinto, E. & Soriano, C. R. (2020). 'Hey, I like your videos, super relate!': Locating sisterhood in an online intimate public on YouTube, *23*(6), 892-907. *Information, Communication, and Society.* doi:10.1080/1369118X.2020.1751864

Cabalquinto, E. C., & Wood-Bradley, G. (2020). Migrant platformed subjectivity: Rethinking the mediation of transnational affective economies via digital connectivity services. *International Journal of Cultural Studies, 23*(5), 787–802. doi:10.1177/1367877920918597

Cabañes, J. V., & Acedera, K. (2012). Of mobile phones and mother-fathers: Calls, text messages, and conjugal power relations in mother-away Filipino families. *New Media & Society, 14*(6), 916–930. doi:10.1177/1461444811435397

Cabañes, J. V. & Collantes, C. (2020). Dating apps as digital flyovers: Mobile media and global intimacies in a postcolonial city. In J. V. Cabañes & L. Uy-Tioco (Eds.), *Mobile media and social intimacies in Asia: Reconfiguring local ties and enacting global relationships* (pp. 97–114). Springer.

Camus, M. (2019). 3rd telco rollout starts moving. *Inquirer Business,* Retrieved 12 January 2020, https://business.inquirer.net/280343/3rd-telco-rollout-starts-moving

Crisostomo, J. (2020). What we do when we #PrayFor: Communicating post humanitarian solidarity through #PrayForMarawi. *Plaridel: A Philippine Journal of Communication, Media, and Society,* 1–34.

Chu, D. (2009). Collective behavior in YouTube: A case study of 'Bus Uncle' online videos. *Asian Journal of Communication, 19*(3), 337–353.

Cunningham, S. (2012). Emergent innovation through the coevolution of informal and formal media economies. *Television and New Media, 13*(5), 415–430. doi:10.1177/1527476412443091

Cunningham, S., & Craig, D. (2017). Being 'really real' on YouTube: Authenticity, community and brand culture in social media entertainment. *Media International Australia, 164*, 71–81.

Cunningham, S., Craig, D., & Silver, J. (2016). YouTube, multichannel networks and the accelerated evolution of the new screen ecology. *Convergence: The International Journal of Research into New Media Technologies, 22*(4), 376–391. doi:10.1177/1354856516641620

David, C. (2013). ICTs in political engagement among youth in the Philippines. *The International Communication Gazette, 75*(3), 322–337.

David, C., Ong, J., & Legara, E. F. T. (2016). Tweeting Supertyphoon Haiyan: Evolving functions of Twitter during and after a disaster event. *PloS one, 11*(3), e0150190. doi:10.1371/journal.pone.0150190

David, E. (2015). Purple-collar labour: Transgender workers and queer value at global call centers in the Philippines. *Gender & Society, 29*(2), 169–194.

De Guzman, O. (2003). Overseas Filipino workers, labour circulation in Southeast Asia, and the (mis)management of overseas migration programs. Retrieved 13 March 2013, http://kyotoreview.cseas.kyoto-u.ac.jp/issue/issue3/article_281.html

Department of Information and Communications Technology. (2017). *National broadband plan: Building infostructures for a digital nation*. Diliman, Quezon City: Department of Information and Communications Technology.

Dobson, A. R. S. (2015). Girls' 'pain memes' on YouTube: The production of pain and femininity on a digital network. In S. Baker, B. Robards, & B. Buttigieg (Eds.), *Youth cultures and subcultures: Australian perspectives* (pp. 173–182). Ashgate Publishing Limited.

Duffy, B. E. (2017). *(Not) getting paid to do what you love: Gender, social media, and aspirational work*. Yale University Press.

Edwards, R., & Tryon, C. (2009). Political video mashups as allegories of citizen empowerment. *First Monday, 14*(10). doi:10.5210/fm.v14i10.2617

Ellwood-Clayton, B. (2005). Texting God: The Lord is my textmate – Folk Catholicism in the cyber Philippines. In K. Nyiri (Ed.), *A sense of place: The global and the local in mobile communication* (pp. 251–265). Passagen.

Fabros, A. (2016). *Outsourceable selves: An ethnography of call center work in a global economy of signs and selves*. Ateneo de Manila University Press.

Flick, U. (2011). *Introducing research methodology*. SAGE Publications Ltd.

Gajjala, R. (2013). *Cyberculture and the subaltern: Weavings of the virtual and real*. Lexington Books.

García-Rapp, F. (2017). Popularity markers on YouTube's attention economy: The case of Bubzbeauty. *Celebrity Studies, 8*(2), 228–245. doi:10.1080/19392397.2016.1242430

García-Rapp, F., & Roca-Cuberes, C. (2017). Being an online celebrity: Norms and expectations of YouTube's beauty community. *First Monday, 22*(7). doi:10.5210/fm.v22i7.7788

Goggin, G. (2011). *Global mobile media*. Routledge.

Guevarra, A. R. (2010). *Marketing dreams, manufacturing heroes: The transnational labour brokering of Filipino workers*. Rutgers University Press.

Hjorth, L. (2011). It's complicated: A case study of personalisation in an age of social and mobile media. *Communication, Politics & Culture, 44*(1), 45–59.

Holmes, S. (2017.) 'My anorexia story': Girls constructing narratives of identity on YouTube, *Cultural Studies, 31*(1), 1–23, doi:10.1080/09502386.2016.1138978

Hutchinson, J. (2017). *Cultural intermediaries: Audience participation in media organisations*. Palgrave Macmillan.

IT & Business Process Association Philippines (IBPAP) (2020). Recalibration of the Philippine IT-BPM industry growth forecasts for 2020–2022. Retrieved 11 March 2021, https://www.ibpap.org/knowledge-hub/research

ITU. (2020). Mobile cellular subscriptions (2001–2019). Retrieved 18 December 2020, https://www.itu.int/en/ITU-D/Statistics/Pages/stat/default.aspx

Jancsary, D., Hollerer, M. and Meyer, R. (2016). Critical analysis of visual and multimodal texts. In R. Wodak and M. Meye (eds.), *Methods of critical discourse studies* (pp. 180–204), 3rd ed.: Sage.

Jenkins, H. (2006). *Convergence culture: Where old and new media collide*. New York University Press.

Jenkins, H., Ford, S., & Green, J. (2013). *Spreadable media: Creating value and meaning in a networked culture*. New York University Press.

Kumar, S. (2016). YouTube nation: Precarity and agency in India's online video scene. *International Journal of Communication, 10*, 5608–5625.

Labour, J. S. J. (2020). Mobile sexuality: Presentations of young Filipinos in dating apps. *Plaridel: A Philippine Journal of Communication, Media, and Society, 17*(1), 247–278.

Lange, P. (2009). Videos of affinity on YouTube. In P. Snickars & P. Vonderau (Eds.), *The YouTube reader* (pp. 70–88). National Library of Sweden.

Lange, P. (2014). Commenting on YouTube rants: Perceptions of inappropriateness or civic engagement? *Journal of Pragmatics: An Interdisciplinary Journal of Language Studies, 73*, 53–65. doi:10.1016/j.pragma.2014.07.004

Lorenzana, J. A. (2016). Mediated recognition: The role of Facebook in identity and social formations of Filipino transnationals in Indian cities. *New Media & Society, 18*(10), pp. 2189–2206. doi:10.1177/1461444816655613

Lorenzana, J.A. (2021). The potency of digital media: Group chats and mediated scandals in the Philippines. *Media International Australia*. doi:10.1177/1329878X21988954

Lorenzana, J. A., & Soriano, C. R. R. (2021). Introduction: The dynamics of digital communication in the Philippines: Legacies and potentials. *Media International Australia, 179*(1), 3–8. doi:10.1177/1329878X211010868

Madianou, M., & Miller, D. (2012). *Migration and new media: Transnational families and polymedia.* Routledge.

Marwick, A. E. (2013). *Status update: Celebrity, publicity, and branding in the social media age.* Yale University Press.

Mateo, J. (2018). Philippines still world's social media capital. *Philippine Star.* Retrieved 19 December 2018, https://www.philstar.com/headlines/2018/02/03/1784052/philippines-still-worlds-social-media-capital-study

McKay, D (2010) On the face of Facebook: Historical images and personhood in Filipino social networking. *History and Anthropology, 21*(4): 483–502.

Mercurio, R. (2019). Philippines among top markets for YouTube. Retrieved 1 November 2019, https://www.philstar.com/business/2019/07/28/1938388/philippines-among-topmarkets-YouTube

Olchondra, R. (2011). YouTube Philippines launched. Retrieved 12 January 2019, https://technology.inquirer.net/5395/YouTube-philippines-launched

Ong, A. (2006). *Neoliberalism as exception mutations in citizenship and sovereignty.* Duke University Press.

Ong, J. (2015). *The poverty of television: The mediation of suffering in class-divided Philippines.* Anthem Press.

Ong, J. & Cabañes, J. (2018). *Architects of networked disinformation: Behind the scenes of troll accounts and fake news production in the Philippines.* Newton Tech4Dev Network. doi:10.7275/2cq4-5396

Ookla (2020). Speedtest global index. Retrieved 14 November 2020, https://www.speedtest.net/global-index

Padios, J. (2018). *A Nation on the line: Call centers as postcolonial predicaments in the Philippines.* Duke University Press.

Paragas, F. (2009). Migrant workers and mobile phones: Technological, temporal, and spatial simultaneity In R. S. Ling & S. Campbell (Eds.), *The reconstruction of space and time: Mobile communication practices* (pp. 39–66). Transaction.

Parreñas, R. S. (2001). *Servants of globalization: Women, migration and domestic work.* Stanford University Press.

Parreñas, R. S. (2008). Transnational fathering: Gendered conflicts, distant disciplining and emotional gaps. *Journal of Ethnic and Migration Studies, 34*(7), 1057–1072. doi:10.1080/13691830802230356

Pertierra, A. (2017). Celebrity politics and televisual melodrama in the age of Duterte. In N. Curato (Ed.), *A Duterte reader: Critical essays on Rodrigo Duterte's early presidency* (pp. 219– 229). Ateneo de Manila University Press.

Pertierra, A.C. (2018). Televisual experiences of poverty and abundance: Entertainment television in the Philippines. *The Australian Journal of Anthropology, 1*(3). doi:10.1111/taja.12261

Pertierra, A. C. (2021). Entertainment publics in the Philippines. *Media International Australia*, *179*(1), 66–79. doi:10.1177/1329878X20985960

Pertierra, R. (2006). *Transforming technologies: Altered selves, mobile phone and Internet use in the Philippines.* De La Salle University Press.

Pertierra, R. (2020). Anthropology and the AlDub nation, entertainment as politics and politics as entertainment. *Philippine studies: Historical & ethnographic viewpoints, 64,* 289–300.

Pertierra, R., Ugarte, E., Pingol, A., Hernandez, J., & Dacanay, N. L. (2002). *TXT-ing selves: Cellphones and Philippine modernity.* De La Salle University Press.

Philippine Statistics Authority (PSA) (2018). *2015/2016 industry profile: Business process outsourcing (first of a series), LabStat updates.* Retrieved 20 December 2020, https://psa.gov.ph/content/20152016-industry-profile-business-process-outsourcing-first-series-0

POEA. (2019). Philippine Overseas Employment Administration, 2015–2016 overseas employment statistics. Retrieved 21 July 2019, www.poea.gov.ph/ofwstat/compendium/2015-2016%20OES%201.pdf

Poell, T., Neiborg, D., & Duffy, B. E. (2022). *Platforms and cultural production.* Polity Press.

Radhakrishnan, S. (2011). *Appropriately Indian: Gender and culture in a new transnational class.* Duke University Press.

Rafael, V. (2000). *White love and other events in Filipino history.* Duke University Press.

Rodriguez, R. M. (2010). *Migrants for export: How the Philippine state brokers to the world.* The University of Minnesota Press.

San Juan, E. (2009). Overseas Filipino workers: The making of an Asian-Pacific diaspora. *The Global South, 3*(2), 99–129.

San Juan, E. (2011). Contemporary global capitalism and the challenge of the Filipino diaspora. *Global Society, 25*(1), 7–27. doi:10.1080/13600826.2010.522983

Senft, T. M. (2008). *Camgirls: Celebrity & community in the age of social networks.* Peter Lang.

Shtern, J., Hill, S., & Chan, D. (2019). Social media influence: Performative authenticity and the relational work of audience commodification in the Philippines. *International Journal of Communication (19328036), 13,* 1939–1958.

Soriano, C.R. (2019). Communicative assemblages of the '*pisonet*' and the translocal context of ICT for the 'have-less': Innovation, inclusion, stratification. *International Journal of Communication, 13*(2019), 4682–4701.

Soriano, C. R., & Cabañes, J. V. (2020a). Between 'world-class work' and 'proletarianized labour': Digital labour imaginaries in the global South. In E. Polson, L. S. Clark, & R. Gajjala (Eds.), *The Routledge companion to media and class* (pp. 213–226). Routledge.

Soriano, C. R., & Cabañes, J. V. (2020b). Entrepreneurial solidarities: Social media collectives and Filipino digital platform workers. *Social Media + Society, 6*(2), 1–11. doi:10.1177/2056305120926484

Soriano, C.R. & Gaw, M.F. (2020). *Banat by: Broadcasting news against newsmakers on YouTube.* Retrieved 30 July 2020, https://www.rappler.com/voices/imho/analysis-banat-by-broadcasting-news-YouTube-against-newsmakers

Soriano, C. R., Hjorth, L., & Davies, H. (2019). Social surveillance and Let's Play: A regional case study of gaming in Manila slum communities. *New Media and Society, 21*(10), 2119–2139. doi:10.1177/1461444819838497

Stockdale, C., & McIntyre, D. (2011). *The ten nations where Facebook rules the Internet.* Retrieved 12 August 2014, http://247wallst.com/technology-3/2011/05/09/the-ten-nations-where-facebook-rules-the-internet/3/

Strangelove, M. (2010). *Watching YouTube: Extraordinary videos by ordinary people.* University of Toronto Press, Scholarly Publishing Division.

Tadiar, N. X. M. (2004). *Fantasy production: Sexual economies and other Philippine consequences for the New World Order.* Hong Kong University Press.

Thompson, N. (2018). *Susan Wojcicki on YouTube's fight against misinformation.* Retrieved 21 May 2022, https://www.wired.com/story/susan-wojcicki-on-youtubes-fight-against-misinformation/

Turner, G. (2009). *Ordinary people and the media: The demotic turn.* SAGE.

UNESCO. (n.d.). Overview of internal migration in the Philippines. Retrieved 22 October 2020, https://bangkok.unesco.org/sites/default/files/assets/article/Social%20and%20Human%20 Sciences/publications/philippines.pdf

Utz, S. & Wolfers, L. (2020). How-to videos on YouTube: The role of the instructor. *Information, Communication & Society,* 1–16. doi:10.1080/1369118X.2020.1804984

Uy-Tioco, C. (2007). Overseas Filipino workers and text messaging: Reinventing transnational mothering. *Continuum, 21*(2), 253–265. doi:10.1080/10304310701269081

Uy-Tioco, C. (2019). 'Good enough' access: Digital inclusion, social stratification, and the reinforce-ment of class in the Philippines. *Journal of Communication Research & Practice 5*(2), 156–171.

Uy-Tioco, C., & Cabalquinto, E. C. (2020). Transnational digital carework: Filipino migrants, family intimacy, and mobile media. In J. V. Cabañes & C. Uy–Tioco (Eds.), *Mobile media and Asian social intimacies,* (pp. 153–170). Springer.

van Dijck, J. (2013). *The culture of connectivity: A critical history of social media.* Oxford University Press.

Visconti, K. (2012). LTE now commercially available in PH. Retrieved 19 September 2014, http:// www.rappler.com/business/11169-lte-now-commercially-available-in-ph

We are Social (2018). Global digital report 2018. Retrieved 29 November 2020, https://digitalreport. wearesocial.com/download

We are Social (2020). Digital 2020: Global digital overview. Retrieved 22 November 2021, https:// wearesocial.com/digital-2020

We are social (2021). Digital 2021. Global overview report. Retrieved 20 November 2021, https:// wearesocial.com/digital-2021

Wilson, C. (2014). The selfiest cities in the world: TIME's definitive ranking. Retrieved 2 Sep-tember 2014, http://time.com/selfies-cities-world-rankings/

Wyatt, S. (2021). Metaphors in critical Internet and digital media studies. *New Media & Society, 23*(2), 406–416. doi:10.1177/1461444820929324

2   Brokering in a Digital Sphere

Abstract

This chapter presents the project's theoretical framework of brokerage as a lens for critically examining how digital narratives, online performativity, and platform-specific strategies reflect how Filipinos navigate the terrain of neoliberal economies and postcolonial legacies on YouTube. The chapter also highlights the four dimensions of digital brokering, including aspirational content, discursive style, credibility building, and platform-specific strategies. In sum, the chapter introduces the dynamics of digital brokerage on YouTube in several key areas of investigation – bleached skin, interracial intimacy, world-class labour, and progressive governance.

Keywords: affective aspiration, brokerage, digital economy, paradoxical reconfiguration, neoliberalism, postcolonialism

This chapter presents "brokerage" as a lens to critically examine how digital narratives, online performativity, and platform-specific strategies reflect how ordinary and marginalised individuals navigate the terrain of neoliberal economies and postcolonial structures in social media platforms such as YouTube. These individuals are afforded with "speaking positions" (Gajjala, 2013) to represent, curate, and broadcast their everyday experiences and harness aspirations and dreams. Yet, making visible these personalised experiences, as shaped by invisible and broader hierarchical structures, reflect anxieties, struggles and constant negotiations in the digital and contemporary society.

This chapter is divided into several sections. First, we discuss the sociological literature on the traditions of brokerage. This approach enables us to engage with the process, politics, and dynamics of brokerage in facilitating social, economic, and political aspirations and interactions. Secondly, we discuss how brokering can be understood in the digital context and highlight how social media influencers enact the process of brokerage. Their presence as "real people" and their performances contribute to affective engagement

Soriano, Cheryll Ruth and Earvin Charles Cabalquinto: *Philippine Digital Cultures: Brokerage Dynamics on YouTube*. Amsterdam: Amsterdam University Press, 2022
DOI: 10.5117/9789463722445_CH02

among audiences, as reflected in user feedback through views, shares, and subscriptions. Further, we situate the discussion in the context of YouTube, noting its key affordances, business model, and data governance mechanisms. This approach allows us to articulate the principles and practices of neoliberal social conditions online. Thirdly, we approach the discussion on digital practices through a postcolonial lens, situating the historical, economic, political, and socio-technological forces that influence performativity in digital spaces. We outline how brokerage takes place in several key areas of investigation – bleached skin, interracial intimacy, world-class labour, and progressive governance. This section is then followed by articulating our proposed conceptual frame of brokering in an online context. We present the four key dimensions in analysing digital brokering in a neoliberal and postcolonial arena – aspirational content, discursive style, credibility building, and platform-specific strategies. These elements enable us to understand how brokering operates by capitalising on "affective aspirations" and producing "paradoxical reconfigurations." Ultimately, this chapter sets the foundation to support and explain our conception of digital brokering, which is applicable to lived realities in the Philippines, with possible resonances for digital cultures elsewhere, as citizens navigate the precarious terrain of neoliberal globalisation and colonial legacies.

Traditions of Brokerage

The concept of brokerage has been used as a critical lens for analysing social divides and mobilisation of social relations. For the former, brokerage has been a central point for enacting social, economic, and political transactions, especially when groups "monopolize goods or information and restrict access to outsiders" (Stovel & Shaw, 2012, p. 140). When information is restricted and not well-distributed, opportunities for brokerage emerge. For the latter, it is understood as the "process of connecting actors in systems of social, economic, or political relations in order to facilitate access to valued resources" (Stovel & Shaw, 2012, p. 141). The broker benefits from the social, economic, and political transactions underlying the brokerage. However, far from being just a simple transaction, brokerage becomes an anchor for people's capacity to imagine a better life and future and how these can be achieved, and therefore facilitates the formation of subjectivities, sociality, and capitals.

The crucial characteristics of brokers are that "(a) they bridge a gap in social structure and (b) they help goods, information, opportunities, or knowledge flow across that gap" (Ibid.). Capitalising on uncertainty and aspiration (Kern & Müller-Böker, 2015, p. 163), for example, property brokers gain value for their work when they are able to pull together crucial information from diverse sources – comparison of real estate rates from different providers, taxation and amortisation options, as well as insurance packages – that would appeal to clients who have difficulty making sense of such information by themselves. By coordinating with multiple entities, brokers effectively put together information into accessible knowledge that becomes marketable to clients when this information is not made readily accessible elsewhere. Here, the process of brokering is enacted by different social actors and institutions within a national and co-located space.

Brokers range from large recruitment firms to local organisations and individual informal agents that are present and visibly entrenched in communities. Their embeddedness in communities strengthens perceptions of their "bias" and "cohesion" (Stovel & Shaw, 2012, p. 142) with those for whom they are brokering. In this framework, bias and cohesion in brokerage pertain to "the extent to which the broker is relationally, socially, or informationally closer to one party than the other, whereas cohesion describes the level of internal solidarity among sets of actors linked to the broker" (pp. 142–143). Bias and cohesion, both constructed through perception, are important in the brokerage process because brokers who are allied with one party more than the other may yield a different result or implication for the brokerage (Ibid, p. 143). Examining the extent to which brokers are viewed to belong to the same community or have their best interest in mind, scholars differentiate between various configurations of brokerage, arguing that subtle imbalances and shifts in the structure of ties affect the brokerage that is made possible. For example, many brokers draw from their past "experiences" or capitalise on local ties to assume a position of relatability and generate trust, allowing them to project the difficulties experienced by an ordinary public and in turn, gain credibility to perform the brokerage role. Other entities may have the same wealth of information and networks that brokers have, but it is the important combination of bringing together dispersed information as well as having in-group identification that a broker's role is legitimated. By coordinating with multiple entities, brokers effectively put together information into accessible knowledge that becomes appealing to clients when this information is not made readily accessible elsewhere.

Aside from the capacity to bring together information, brokers generate value by being connected to actors who are unconnected (Burt, 2007), but also in facilitating new connections. As such, the brokerage process is

considered to be "one of a small number of mechanisms by which discon-
nected or isolated individuals (or groups) can interact economically, politically, and socially" (Stovel & Shaw, 2012, p. 140). In this way, brokers mediate
multiple relationships among actors too. This implies that understanding
brokerage requires examining not just the broader structures underscoring
them, but also the micro-processes and micro-level relations among parties.
It is also important to examine the dual aspect of brokerage. On one hand,
brokerage has the capacity to ease social interaction, enhance economic
activity, and facilitate political development. On the other hand, brokerage
can breed exploitation, the pursuit of personal profit, corruption, and the
accumulation of power; through these and other processes, brokerage can
exacerbate existing inequalities (Stovel & Shaw, 2012, p. 140). While in past
literature brokers have focused largely on the accumulation of finance
capital, there is also a long tradition of brokers accumulating other forms of
capital (McKay & Perez, 2019). Foremost is the accumulation of social and
cultural capital as brokers derive legitimacy and social influence through
the brokerage process.

The concept of brokering has been actively deployed in a labour migration context. Several scholars have highlighted the interaction between
social actors, transnational institutions, and infrastructures that facilitate
transactions, flows of people, and capital access (Lindquist, Xiang, & Yeoh,
2012; Shrestha & Yeoh, 2018). Brokers transact for workers from low-income
and rural communities who lack consolidated information on how to access
overseas labour opportunities, as well as know-how for strategically marketing their skills and capabilities to appear eligible for such jobs (Lindquist et
al., 2012; Rodriguez, 2010; Shrestha & Yeoh, 2018). Their work is crucial if we
think about the humongous challenge of navigating Philippine bureaucracy.
According to a government report, for example, aspiring overseas workers
and seafarers need around "2–3 months to obtain 70–73 signatures from
10–14 different government agencies" just to be able to secure the clearance to
work overseas (Information Technology and e-Commerce Council Roadmap,
2003: np). For an aspiring worker without access to informational resources,
the service offered by migration brokers is indispensable for helping aspirants understand complex labour migration procedures, simplifying the
complicated steps and requirements, making sense of the different agencies
and their documentary requirements, and navigating local and foreign
bureaucracies.

The persistence of uneven economic conditions also influences brokerage.
This is shown in the context of development aid, involving the complex
distribution of resources from the global North to Global South (Jensen,

2018; McKay & Perez, 2019; J. Ong & Combinido, 2017; Saban, 2015). Studies have shown that transfers of information and money between donors and beneficiaries are brokered by citizen aid brokers, implying that trust relationships are developed not between donors and beneficiaries, but between citizen aid intermediaries and beneficiaries. Although brokers intermediate for established organisations and channels, they operate through highly personalised and informal processes. In the brokerage process, diverse channels are utilised by global and national organisations to facilitate the flows of information and deliver aid (McKay & Perez, 2019). By setting up new sources of support and information, brokers thrive in a country where government support is unable to meet the needs of recalcitrant publics. Notably, their operation in delivering aid allows them to establish control over local publics by identifying a list of whom they have "brokered for," thus, "reinvigorating inactive obligations from previous exchanges in order to create a constituency for their activities" (McKay & Perez, 2019, p. 1906). While brokers can be resented by clients, either by the brokerage fees they require or the influence that they wield, they are tolerated because they have connections and resources that are useful, if not indispensable, and which can be readily tapped. And while brokers may deliver on their promises, they also operate in ways that put their own interests first, or at least along with those of their clients.

Brokerage in the Philippines

The Philippines is an important case study to understand brokerage. As presented in Chapter 1, the term brokerage has been applied to characterise the Philippine state in the process of actively pushing and even "manufacturing" its citizens for overseas labour migration through its labour export policy. Rodriguez (2010) defines brokerage as "...a neoliberal strategy that is comprised of institutional and discursive practices through which the Philippine state mobilises its citizens and sends them abroad to work for employers throughout the world while generating a 'profit' from the remittances that migrants send back to their families and loved ones remaining in the Philippines" (p. x). Guevarra (2010), on the other hand, articulates brokerage as a mechanism for image making of labour export. For instance, Filipino workers are often framed by the State as "ideal workers" because of their competitive skills in using the English language, their self-sacrificing and subservient character, as well as their hardworking attitude (Soriano & Cabanes, 2020a). Although the state has staunchly endorsed its labour export policy, there is a general shortcoming in terms of informational and practical support for aspiring migrant workers. Non-state brokers, such as labour migration agents and recruiters, fill this gap. According to Shrestha and Yeoh (2018), it is the brokers' wealth of information on local and global labour migration processes that legitimise their work, assist and condition Filipino migrant workers with the promise that after years of overseas work, they would eventually be able to recuperate the brokerage fees. The brokers then profit from using their skills, knowledge, and networks crucial for making sense of foreign and local bureaucracies (Lindquist et al., 2012). Ultimately, brokerage mobilises transactions between employers and workers, clients and customers, or even donors and beneficiaries, which process facilitates access and exchange of capital.

Brokerage is not solely tied to labour exportation. It has also been deployed in entangling racialised, gendered and classed bodies in the beauty and marriage migration industry. In brokering beauty ideals, key figures such as beauty scouts, talent managers, models, and celebrities act as brokers in conveying ideal femininity (Baldo-Cubelo, 2015; Rondilla & Spickard, 2007) and masculinity (Lasco & Hardon, 2020) subjectivities. Prevalent is the functioning of recruiters – beauty scouts and talent managers – who perform the role of identifying the next "superstar" or "surprise beauties or bodies" from the countryside or low-income communities and offering them celebrity deals or an opportunity to compete in national beauty pageants. The discovery of a "celebrity" and the developments surrounding that dis-covery – from which community to what contract is eventually clinched – is covered in Philippine media. One prominent case is *Isko Moreno*, who was recruited by talent show host German Moreno from the slums of Tondo for his looks, and who has become a successful actor and now also the mayor of the city of Manila. These recruiters, capitalising on aspirations, facilitate a bridge between an ordinary Filipino who possesses particular ideals of beauty – often Caucasian standards of having fair skin, well-chiselled nose, or deep-set eyes – and his or her aspirations for recognition and celebrity stardom. Becoming a model or an entertainment celebrity is unthinkable for an ordinary lass or lad from the countryside, and brokers put together crucial information – how to dress, how to walk, how to use make-up, how to style the hair, how to talk – and connect them to industry networks. The broker, who works as an intermediary between media companies and individuals, then obtains commissions from contracts that the new discovery makes. In Philippine society, beyond direct beauty brokering by beauty scouts and talent managers, representations of attractiveness and desirability are further mobilised through media representation of these "discovery" processes and further echoed in the aggressive promotion of beauty and whitening products everyday by beauty companies and their celebrity endorsers (David, 2013; Glenn, 2008, 2009; Rondilla, 2009). Several scholars have highlighted that aspiring for and embodying "whiteness" or having a fair-skin – a trope recurring in Philippine advertisements and articulated through the preference for fair-skinned models or protagonists, is linked to racial and colonialist hierarchy in the Philippines (Rafael, 2000; Rondilla, 2012). In a contemporary digital context, we can see the emergence of YouTubers who not only present beauty standards of whiteness as a ticket to recognition, attention, and ultimately success, but practical ways of achieving such beauty standards that an ordinary Filipino can adopt. These themes are presented in Chapter 3.

In the context of brokering marriage migration, studies show that some Filipino women meet their foreign partners by subscribing to matchmaking services or using online tools (Constable, 2003; Tolentino, 1996). Further, brokerage is facilitated by catalogues or websites (Constable, 2003; Tolentino, 1996) that provide detailed information or offer possibilities for geographically dispersed individuals to meet, interact, and even marry. In a sense, matchmakers typically coordinate and pair a woman with a man, with their correspondences occurring via letters or electronic messages (Constable, 2003). Meanwhile, scholars have argued that interracial pairing demonstrates a form of commodification of intimacy, wherein relationships are forged and further enhanced in transactional or economic terms (Constable, 2009). A cohort of studies has shown how representations of interracial relationships in the media often stereotype Filipino women as "mail-order" brides (Gonzalez and Rodriguez, 2003; Robinson, 1996; Saroca 2002, 2007). As a result, Filipino women who are in these types of relationships or marriage arrangements are stigmatised, discriminated and even abused (Aquino, 2018; Laforteza, 2016; Saroca, 2007). Meanwhile, Filipino scholar Rolando Tolentino (1996) argues that the country's colonial past and neoliberal policies fuel the mobilisation of Filipino women for marriage migration. In contrast, Constable (2003) contends that Filipino women who engage with matchmaking networks and channels have diverse reasons and agentic capacities for marriage migration. Filipino women also deploy "countererotics" or ways that shun stereotypes through online narratives (Constable, 2012), such as visibilising their caring, loving and domestic persona. These points on the complexity of brokering intimate partners on YouTube will be further discussed in Chapter 4.

As we will discuss in Chapter 5, labour arrangements can also be mediated by emerging brokers on YouTube. Tracing the labour brokerage process from

Philippine historical and colonial roots, we note the crucial role played by *cabos* in negotiating and supplying labour for various kinds of industries during the Spanish colonial system. In this context when Filipinos did not have the right to own land, transacting with brokers was the only way for many peasants to land a job as seasonal agricultural workers (IBON Foundation, 2017; Kapunan, 1991; Silarde, 2020; Soriano, 2021). Where Filipinos had no choice but to work with agricultural contractors, *cabos* performed the role of middlemen and acted as negotiators between workers and contractors, determining how much workers would be paid for a specific scope of work or in helping workers determine where work was available (Kapunan, 1991, p. 326). This labour arrangement can be extended in the digital realm, as we highlight the role of platform labour influencers emerging on YouTube.

Building on some of the themes we presented, we ask, what conditions exactly underscore Philippine society and culture that makes it conducive for brokerage? We argue that the process of brokerage is not devoid of politics. As such, we articulate brokerage as working well within pervasive "patron-client relations" (Nowak & Snyder, 1974) that underscore how brokers obtain the legitimacy to bridge ordinary people and their aspiration to access resources that are often held by elites; in turn, elites benefit from this brokerage process by retaining their control over surplus (Nowak & Snyder, 1974, p. 23). Yet, elites can appear distant from the masses, and Nowak and Snyder (1974) point to the role of political intermediaries, often emerging from the middle classes, to bridge the national elite to the masses. Here we emphasise the ambivalent reputation of brokers where they frame themselves as "political or community mobilisers" – focusing on their contribution to direct needs such as people's livelihood opportunities as well as to broader aspirations such as economic and political development (Kern & Müller-Böker, 2015, p. 18) while performing reputation management and advancing political agendas.

This implies that brokerage is also inscribed in political aspirations. As we will discuss further in Chapter 6, where politics is about mobilising interests, brokerage is about influencing and controlling the flow of information and resources to allow parties to achieve their goals. Politicians engage brokers to bridge them with ordinary constituents and obtain votes. On the other hand, brokers also promise constituents resources, a political vision, or an illusion of closeness to the politician in exchange for a vote. Functioning to bridge voters with political candidates (and political aspirations) through "personalised patronage," brokers obtain influence by directly mobilising people's interests and aspirations, making them inconspicuous but crucial players in the political process. Yet, with politics becoming more embedded in social media, specialised patron-client relationships have emerged in more functionally specific forms such as political machines and disinformation architects (J. Ong & Cabañes, 2018; J. Ong, Tapsell, & Curato, 2019), which allows for the re-enactment of political brokerage in digital form.

Ultimately, this book focuses on the conditions and dynamics that allow non-state, informal, and individual brokers to thrive in a digital environment, particularly on a video-based platform such as YouTube, which has also been increasingly used by Filipinos for diverse personal, political, and even commercial purposes. In digital spaces, figures such as influencers, micro-celebrities, and cultural intermediaries capitalise on their skills, knowledge, expertise, cultural taste, social networks, experience, technological literacies, and a range of capitals to craft and circulate relatable, intimate, personalised and affective content. Like a broker, the YouTuber weaves together a variety of information that viewers seek, and it is through the capacity to present this organised information and create networks between people and between people and other actors that his or her labour gains value. Social media can produce novel forms of brokerage through the circulation of images from which aspirations can be anchored (McKay & Perez, 2019). We turn to the next section to discuss this point further.

Influencers as Brokers

The literature on influencer culture abounds (Abidin, 2018; Duffy, 2017; Marwick, 2013, 2015), but the examination of how influencers broker political, economic and cultural aspirations in the Global South has not been fully explored. In the first instance, influencers pertain to social media users with niche audiences and following and who employ creative strategies adopted from traditional celebrity culture to maintain attention and cultivate relationships with those audiences (Abidin, 2015; Marwick, 2013). Further, they amass a smaller scale of followers compared to traditional celebrities, but their strategies can forge a loyal following on any social media platform (Khamis, Ang, & Welling, 2017). As Hearn & Schoenhoff (2016, p. 194) explain, these individuals amass "celebrity capital" by cultivating as much attention as possible and crafting an authentic and relatable "personal brand" via social networks, which they can also subsequently monetise. They often create a persona to differentiate oneself and the content that they create (Abidin, 2015) while using "personal sensations" to establish a connection with the audience and target emotions (Malefyt & Morais 2012, p. 62).

Influencers can be characterised as cultural intermediaries through their participation in the production of symbolic and cultural goods (Bourdieu, 1984). By being "in between" producers and consumers of goods, they deploy strategies for generating capital and relatability while also being translators of tastes, language, norms, rules, and regulatory frameworks between the mainstream institutions and audience stakeholder groups (Hutchinson, 2017, p. 21). Yet, they also establish some "distance" from the industry or the mainstream by creating their own contents (Negus, 2002; Lewis, 2020, 2019) while connecting media institutions and audiences through co-collaboration or co-creation of contents (Hutchinson, 2017).

Influencers perform their role similarly to cultural intermediaries through branding strategies (Duffy, 2017; Abidin, 2018). Diverse narratives, tactics, and collaborative work allow them to enact their role as "capital translators" that add value to media goods and services (Hutchinson, 2017). Notably, branding strategies fuel the operations of a neoliberal economy, shifting the locus from institutions to individuals positioned to be self-driven, innovative, and entrepreneurial in online spaces (Duffy, 2017; Marwick, 2013). One way media creators render their content attractive and of value is by engaging with the platform's features (Burgess & Green, 2009) – visuals, sounds, narrative style, self-recommendations, comment features, among others. This enables the rise of ordinary people to take on the role of "experts" through tutorials and "explainer" videos that pull together diverse informational content and blend information, interpretation, and entertainment in their videos and channels, even as they also sometimes echo mainstream narratives (Hou, 2018; Utz & Wolfers, 2020). As Lewis (2020) has argued, "micro-celebrity practices are not only a business strategy but also a political stance that positions them as more credible than mainstream media" (p. 1).

The "influencer" as a concept is largely understood in niche-like environments such as fashion, entertainment, blog, travel and the like, and often in the context of affluent societies. Similarly, the operation of brokers has often been studied in the context of more traditional face-to-face transactions. However, the relationship between influencers and brokerage is not well-established, but we can see intersections in their strategies and how their work attracts value. Drawing from our earlier discussions of brokerage traditions, we can see brokers' crucial role as bringing together dispersed information into knowledge as well facilitating connections among disparate and previously unconnected actors which they then use to build profitable social transactions. The broker intermediates for a client and an aspiration. So does an influencer, earning clicks and likes for effectively selling a lifestyle or an idea deemed of value to their target publics by satiating specific aspirations.

Like influencers, brokers adopt a relatable persona (i.e. of a low-profile expert within a field of interest – whether this is migration, stock exchange, real estate, humanitarian assistance, or car dealership), crafts and maintains "communicative intimacies" (Abidin, 2015) to convey this expertise, while maintaining a loyal audience. Community-building also serves as an important anchor for the communicative strategies enacted by social media influencers (Garcia-Rapp, 2017; Senft, 2008; Abidin, 2015). When an influencer has established an expertise or persona, he or she can build on this persona by maintaining and targeting the communication to initial subscribers, actively referring and addressing them as part of his or her community (Hou, 2018) and with the constant intention of growing this community to facilitate a more organic exchange of content and maintenance of income stream.

As we will discuss in the succeeding chapters, like influencers, brokers adopt strategies to perform their brokerage role on YouTube by attuning their performance to the platform's range of affordances. However, a YouTuber needs to have the necessary skills for brokerage – digital proficiency, technical skills for production and editing, and importantly, cultural know-how of the YouTube's language and features and how to optimise these. How influencers are conscious of the platform's vernacular and apply these in their presentation and engagement tactics crystallise their brokerage role. With the creation of previously unavailable linkages (i.e. between information and between people) being one of the key roles of brokers, the presentation of relatable knowledge and enacting a community through the engagement of social media affordances illustrates the tight relationship between influencer practices and brokerage in this contemporary digital environment.

YouTube as a Socio-Technical Broker

YouTube's role in the information ecosystem has shifted from catering to lifestyle videos into enabling and hosting "epistemic communities" where creators can make claims over knowledge (Utz & Wolfers, 2020). This development collides with participatory and do-it-yourself (DIY) culture on social media, where creators build content and share this with communities that then translate this knowledge into "collective intelligence" (Jenkins, 2006). This DIY culture of content creation departs from traditional expert cultures of knowledge creation and sharing underscored by gatekeepers and editors of content, bringing into fore new social imaginaries of YouTube as a platform (Hou, 2018). On YouTube, the audience and the creators are in a constant gravitational force and continuously overlapping, challenging institutionalised forms of knowledge, with even the possibility of subordinating expertise (Marchal & Hu, 2020).

YouTube facilitates "social learning" (Utz & Wolfers, 2020) in a multimedia/infotainment format that makes information palatable. The richness of visual and moving image accompanied by music makes the conveyance of information more affective and dynamic. This is also supported by the evolution of YouTube into a "hybrid cultural–commercial space" (Lobato, 2016, p. 357; Arthurs, Drakopoulou & Gandini, 2018) where various forms of serious content become entangled with commercial or entertainment styles of presentation. The interplay of visual and auditory styles contributes positively to the narrative that can then trigger multiple audience engagement such as subscribing, viewing, liking, and commenting. Specific to brokerage, YouTube carries with it an inclusive, cultural characteristic that makes it a connective force not only between disparate forms of information, but also between disparate parties. In analysing brokerage on YouTube, we resurface the connectedness of communication systems with social systems, reflecting how technological features shape social relations, but also how social contexts shape technology. To do this, we first examine the platform's governance mechanisms and affordances and then move on to how users appropriate these affordances. YouTube's platform governance mechanisms and algorithms "do things" by embodying a "command structure" (Geofey, 2008, p. 17) that mediate, augment, and condition our most trivial everyday experiences as these intersect with sociality (Bucher, 2016, p. 84). We discuss YouTube's affordances according to the following dimensions: regimes of visibility, ordering and relevance, and curation of content.

One key affordance is the facilitation of "regimes of visibility" (Bucher, 2012) and YouTube performs a powerful role in visibilising content creators and curating their content for interested audiences. While media of all kinds function to make visible aspects of the social world by extending seeing and sensing, they also determine our regimes of visibility, with platforms algorithmically modulating our capacity to achieve social attention or alternatively be rendered "invisible" (Bucher, 2012). Here, we mean not just the visibility of a single video, but of a YouTuber's series of videos organised in a Channel, as well as the platform's affordances of allowing YouTubers to categorise their content to facilitate their visibility upon search. The platform's capacity to make visible a set of related content in the form of videos, video descriptions, comments, and so on can create a composite of information that serve as persuasive narratives. Further, the dynamics of discovery are reworked as algorithms allow culture to "find" us instead of us looking for it (Lash, 2007). For example, Rieder and colleagues (2018) found that fresh news and trending issues were more likely to be surfaced on YouTube over other content. The visibility of YouTubers and their content can be facilitated amid porous governance mechanisms over who creates content and what information they share or knowledge claims they might make (Bishop, 2019). This algorithmic visibility is a crucial ingredient for brokerage to take place and can be understood as a part of "aspirational labour" as YouTubers build careers and following via social media presence and visibility (Duffy, 2017).

Connected to the notion of visibility is the process of "ordering content" and giving relevance to certain objects in the way they are sequenced in the search results and recommended videos, implicitly rendering credibility and legitimacy to certain actors (and their narratives) over others (Gillespie, 2016). Beyond making content visible, the platform's "ranking culture" (Rieder et al., 2018, p. 52) determines the level of value audiences give to them, with studies finding that while a broad range of content are made "visible," attention is directed to the content that platforms privilege in terms of higher rank or order (Amoore, 2011). Studies also found that commercial logics often determine their rubric of relevance, privileging trends and novel forms of cultures (Gillespie, 2014).

Beyond visibility and ordering of content, perhaps an important affordance of social media platforms that also emerge from platform user-interaction is how the platform "curates" content for particular users. Curation works by categorising and filtering content according to the user's views and preferences, and with computer algorithms having the capacity to infer categories of identity of users based on their consumption behaviour. The extensive surveillance of users enables platforms to construct "algorithmic identities" to personalise users' experiences (Cheney-Lippold, 2011). Considered to function as technological platforms' soft biopolitics and biopower, this curation of content based on a user's algorithmic identity to suit a user's tastes can be considered political practice. Cohn (2019) argues, for example, that the creation of these personalised pathways to discovering specific content also exposes a user to a curated narrative. At the extreme level, research has associated platform curation work toward political polarisation (Jamieson, 2018) and even radicalisation (Fisher & Taub, 2019), where users are targeted by platforms, pushing particular content to them on the basis of their political inclinations, embedding them further in their political bubbles, and thereby relegating them as political subjects of the platforms.

This personalisation of content is highly connected to the brokerage process, especially as brokerage processes thrive in personalised affective narratives linked to people's personal aspirations, interests, and inclinations (Stovel & Shaw, 2012). However, users as well as creators also have a way to circumvent or reinforce the working of algorithms through specific strategies anchored on "algorithmic literacy," where basic knowledge of how filtering mechanisms and design choices work can lead users to "bursting" these bubbles (Reviglio & Agosti, 2020, p. 7) or creators using influencer strategies to both optimise and game the platform's logics.

This social-technical process underlying the platform's capacity to organise visibility, ordering, and curation of content for specific audiences configure the production of culture, economy, and politics and likewise continually shape the strategies and tactics of content producers. Platforms also update and shift their algorithms, shaping what is seen and engaged with without users' consent or knowledge (Bucher, 2018; Bishop, 2019). Research on algorithmic knowledge production investigates "technical" sites, such as recommender systems and tech organisations, to examine how algorithms are constructed, operate, and are continually reconfigured (e.g. Pascuale, 2015; Reviglio & Agosti, 2020).

Beyond the working of the algorithm, it is imperative to explore YouTube's affordances as these intersect with human interaction. While algorithms make content visible, categorise and organise this content, content that drives user engagement depends on how YouTubers as brokers use the platform's other features (Burgess & Green, 2009). Ultimately, the book aims to show that YouTube facilitates the brokerage work of influencers while also assuming a brokerage role through the governance of content between producers and consumers. By brokering digital transactions, not only does YouTube as a platform benefit financially, but it also facilitates the brokerage of social transactions that reinforce the principles of economic globalisation. We now turn to discussing brokerage as moulded by socio-historically situated influences.

Digital Brokering of Aspirations in a Postcolonial Context

Many studies on influencer culture have focussed on neoliberal practices in online spaces (Abidin, 2018; Duffy, 2017; Marwick, 2013). To date, few studies have situated their investigation of influencer practices within the conditions created by colonial hierarchies (Gajjala, 2013; Nakamura, 2013; Sobande, 2017, 2019). This book attends to this gap by situating the principle and practice of brokerage in a digital environment that is deeply enmeshed within neoliberal and postcolonial structures. To elucidate this contribution, we highlight the different aspects of digital brokering as simultaneously shaped by economic globalisation and postcolonial conditions that lead to *affective aspirations* and *paradoxical reconfigurations*.

As we have discussed in the earlier sections, the mediation of aspirations is key to brokerage. In traditional forms of brokerage, aspirations range from obtaining good value property to accessing employment opportunities and social welfare services to obtaining resources in times of crisis. In the digital space, aspirations are anchored on transformations of lifestyle, upskilling, community building, and capital access (Abidin, 2018; Berryman & Kavka, 2017; Glatt, 2017). For viewers, the connection to aspirational content lies in what Kavka (2008) calls, "situated identification." By examining how affect is generated through the performativity of reality TV stars, Kavka (2008) notes that a viewer accesses the performance of reality by real people and relate to such "real" performativity. In this case, the presence of ordinary people on a mediated screen generates affect through the intimacy of consumption.

We develop our understanding of brokering in digital spaces through the notion that a web of affective content and networks facilitate proximity between the on-screen talent and the viewer (Kavka, 2008). We extend this approach that "aspirations" become palatable most especially when ordinary people aspire, act on, and display the achievement of their personal journeys, relationships, and goals. Through a mediated selfhood enacted through images, texts and audio clips (Kavka, 2008), aspirations conveyed by ordinary people on screen accompanied by demonstrations of how these can be actualised and reinforced through audience engagements allow these to obtain a political dimension. In a way, the aspirations presented by the performer "tickle" aspirations for viewers, despite these aspirations being anchored on the realities of inclusion and exclusion in imagined worlds (Appadurai, 1996). As such, these aspirations may remain a fantasy especially for those who do not have the ability and resources to actualise the same aspiration or living condition. In this regard, we argue that emerging in digital spaces are what we refer to as "affective aspirations," a dreamwork mobilised through intimate contents, performances, and strategies that engage viewers who consume content only to live vicariously through a mediated persona's world as a result of one's uneven access to a range of capitals. In this case, affective aspirations essentially become the bind between the mediated persona and the viewer who is aspiring what the mediated persona does. Further, sustained connections are achieved through the visibility of ordinary individuals on screen, who also curates and presents a range of information and tactics through which viewers can consume to achieve the mediated persona's status amid the precariousness of everyday living.

We approach affective aspirations as tied to colonial legacies. We take coloniality to mean "long-standing patterns of power that emerged as a result of colonialism, but that define politics, culture, labour, intersubjective relations, and knowledge production well beyond the strict limits of colonial administrations" (Casilli, 2017, p. 31). Postcolonialism implies that "people's experiences are continually shaped by the unequal hierarchies established by imperialism," but which is continually reconstructed through various interpersonal, national and global processes (Fujita-Rony, 2010, p. 3). Through a post-structuralist lens, implications of coloniality reflect in sets of socially shared representational assemblages and practices that function as contingent anchor points in terms of emerging relationships of individuals to their conditions of existence (Gajjala, 2013; Loomba, 1998). Building on this perspective, mediated spaces can produce acceptable and marketable representational assemblages that are entangled with neo-colonial structures (Jamerson, 2019).

Affective aspirations showcase the raw, mundane, and intimate -- a form of calibrated amateurism (Abidin, 2017) that appears to be entangled with macro socio-historical relations. But we argue that such content reflects consciousness governed by colonial discourses, such as through media representations and interactions (Bhabha, 1994). Our conception of aspirations builds on the work of Stoler and Bond's scholarship on past and present imperial formations, arguing that "the empire works in the everyday" (2006, p. 101). They called on scholars to emphasise the links between the macro and micro in our analyses of postcoloniality by identifying "...structured imperial predicaments by tracing them through the durabilities of duress in the subsoil of affective landscapes, in the weight of memory, in the manoeuvres around the intimate management of people's lives" (2006, p. 95). This point on affective landscapes echoes Bhaba's (1994) contention on the strategies of hierarchisation and marginalisation of colonial cities. In a way, a networked environment represents this reality in the age of modernity.

We argue that the production of affective aspirations can be understood as a key strategy for a mediated persona to connect with the viewers and sustain engagement. As affect circulates in networked environments, visibility and performativity of aspirations can trigger desires to embody the actions, mobilities and lifestyle performed by the mediated persona – the content creator. In digital spaces, the desires for transformation that are being "sold" by the content creator become a bait that attracts audiences to keep coming back and act on the prescribed action and achieve change. In a postcolonial context, the concept of mimicry applies (Bhabha, 1994), wherein a colonised aspires for the position of the coloniser. We contend, in a digital space, the formerly colonised – a content creator with internalised colonialist systems then become followed and emulated by viewers who likewise embody these postcolonial aspirations. However, both the performer and the viewer – who attempt to mimic and embody "colonial positionality and legacies" – may feel ambivalent about such transformations as a resemblance. As Bhabha (1994) notes, "white but not quite." As such, the quest for achieving or negotiating similarity becomes the driving force for continued engagement and actions, and which can reinforce the inferior status of the colonised because of its unwavering desire to be white and dominant (Bhabha, 1994).

In connecting our analysis of brokerage in digital spaces to the broader structure of economic globalisation and postcoloniality, we are inspired by the work of Radhika Gajjala (2013) on micro-transactional and digital practices in Indian society. For Gajjala (2013), the advent of digital communication technologies and online channels have stirred a paradoxical case of empowerment for previously colonised people. On the one hand, they are integrated into digital circuits that activate exchange and sales in global markets. On the other hand, they embody "technocultural agency" in varying degrees depending on socio-economic hierarchies, and asymmetrical capacities to negotiate online and offline, and global and local spheres (Gajjala, 2013).

This perceivable condition being projected by YouTubers and their audiences seems to echo Elisa Oreglia and Rich Ling's notion of "digital imagination." Applied in the context of digital technology use among non-techno-elites in the Global South, digital imagination is described as "the process by which individuals within a society develop an understanding of the potentials, the limitations, and eventually the threats of digital technology" (2018, p. 571). Like Oreglia and Ling's non-techno-elites who grapple with the promise of these new technologies, YouTubers similarly conduct a "mental matching" to envision the possibility of achieving their aspirations (2018, p. 571), which are then projected by other YouTubers. This leads us to argue that YouTube creates a state of mind wherein audiences sustain a global vision or image of possibility whether or not they actually benefit from its promises just yet. Drawing from the notion of the context-specific ways in which individuals utilise imagination and draw from their own socio-cultural positions as a way to make sense of the world despite having only a limited set of resources (Oreglia & Ling, 2018), we refer to the state of mind being channelled by YouTubers as "paradoxical reconfigurations."

We argue that the current digital landscape of influencer cultures operates as a space for reinforcing and negotiating historically situated structures and hierarchies. Certainly, content creators, who come from diverse income classes including those from marginalised communities, are utilising digital technologies and online content to brand themselves as mobilising above their social positions. These digital practices serve to renegotiate, transform, and reinvent the legacies and lived experiences of neoliberalism and postcoloniality. Yet, their embeddedness in the digital sphere does not necessarily change the structures of hierarchy and marginalisation, but instead facilitate the possibilities of living within neoliberal and postcolonial conditions. In this regard, we coin the term "paradoxical reconfigurations" to articulate how transformations in digital spheres that enable modalities for self-representation, expressions of aspiring, and performances of achieving simultaneously demonstrate agentic possibilities in contemporary modernity while replicating the visions, ideology, and practices of neoliberalism and colonial legacies. As to be presented in the next section, the brokerage and embodiment of affective aspirations delivers these paradoxical reconfigurations.

Brokering Filipino Aspirations

We locate the formation of affective aspirations of Filipinos as existing in tension with established institutionalised knowledge imbricated in historical structures of knowledge production, rooted in Philippine history and geography of modernity (Shome & Hegde, 2002). We also look at aspirations for the nation and recognise that the aspirations – whether individual or national – extend beyond the nation. This entails "geopoliticizing the nation and locating it in larger (and unequal) histories and geographies of global power and culture" (Shome & Hegde, 2002, p. 253). By placing the analysis within a broader socio-historical context, we will show YouTube to be an embodied practice transcending both physical and sociocultural distance and empowered by mediated images and discourses, as well as colonial imaginaries about political, economic, social, or cultural mobility through brokerage. The mediated relations between YouTube, content creators and their audiences allow Filipinos to collectively envision their world, as well as their nation, and the possibilities in between.

As shown in Chapter 1, three centuries of Spanish colonial rule and half a century under American rule have left lasting "legacies" that shaped Filipino consciousness and aspiration. Following its independence in 1946, the country also remained an important ally of the United States for political, economic and military reasons, and which succeeding Philippine political governments embraced (Fujita-Rony, 2010, p. 3). This colonial past and present has paved the way for the Filipino's hierarchical racial affinity (Rafael, 2000) and transnational identities (Fujita-Rony, 2010), where Western political and economic systems are often perceived to be more superior, or where migration to these nations is deemed to bring the promise of class mobility entangled with a multitude of indirect social advantages, such as the possibility of "living a better life" or the promise of enriched social influence at home (Cabalquinto & Soriano, 2020).

As we examine the brokerage of affective aspirations on YouTube in the succeeding chapters, we focus on the representational politics and paradoxical outcomes that are deeply ingrained in imperial relations and neoliberal market forces. To do this, the book presents four areas of investigations: bleached skin, interracial intimacies, world-class work, and progressive governance. These topics have been selected to generate insights on how the digital space – YouTube, in particular – is becoming a repository of affective aspirations and paradoxical configurations that tend to perpetuate cultural, economic, and political power structures.

Bleached skin

Scholars have argued that aspiring for a white or fair skin is linked to notions of selfhood underscored by Philippine colonial legacies (David, 2013; Rafael, 2000; Rondilla, 2012). As we have shown in Chapter 1, ideologies of whiteness or a "skin-colour hierarchy" in the Philippines instilled the idea that Spanish colonial masters were the superior race, and this hierarchy is reinforced by creating a division between Filipinos and Filipinos with Spanish bloodlines (Camba, 2012). When Spanish men reproduced with Filipinas, their children and their future generation would be called *mestisos/as* who enjoyed privileged positions associated with economic wealth, political influence, and cultural hegemony while pure Filipinos (*indios*) were more likely to continue to be less educated and have less economic opportunities (Rafael, 2000, p. 165). To date, mixed-race figures embodying the image of the mestiso/a remain privileged and idolised in contemporary Philippine society as celebrities, beauty queens, newscasters, politicians, executives, and so forth (Mendoza, 2014), and they are celebrated in advertisements and beauty pageants that promote standards of Filipina beauty (Baldo-Cubelo, 2015; Rondilla, 2009), thereby reviving what Rafael called, "mestisa envy" (Rafael, 2020, p. 165). In this book, we focus on whitening the underarm, a part of the body which is often framed by beauty and cosmetic brands as a female bodily issue that is abhorred and needing repair. Colonialist bodily transformation is then articulated as a form of "objectified cultural whitening" or the use of products to enact shared aspirations on whiteness (Arnado, 2019). In Chapter 3, we showcase how neoliberal and colonial influences shape the orchestration of embodying a white and smooth underarm.

Interracial intimacy

Colonial influences also manifest in aspiring for a Caucasian partner. This point shows an institutionalised form of cultural whitening wherein marriage provides the mobility and pathway for mixed partnership or marriage (Arnado, 2019). We highlight that the institutionalised dimensions of interracial and intimate aspirations are entwined with neoliberal and postcolonial conditions. In the context of neoliberalism, scholars have presented the historicity of Filipino women's overseas mobility because of the lack of access to job opportunities and social welfare services in the Philippines (Constable, 2006; Tolentino, 1996). In other cases, this mobility is accessible when a white man marries a Filipino woman, which pairing occurs in online spaces (Gonzalez & Rodriguez, 2003). This has led to the popularisation of the term "mail-order bride," a concept that has also been criticised by some scholars, such as Constable (2003) and Saroca (2002) for generalising Filipino women as lacking agency and capacity for decision making. This cultural logic of desire for a white partner is produced through colonialist discourses, fashioned out of a specific historical relationship between the United States and the colony, the Philippines, driven by American popular media content and local media content that echo the same ideals, as well as everyday traditions and practices (Tolentino, 1996). Further, the Philippine colonial past has paved the way for the Filipino's hierarchical racial affinity (Rafael, 2000). This translates to the aspiration for intimate relations leading to marriage with white, foreign nationals, implying promises of class mobility entangled with a multitude of indirect social advantages, such as achieving stability for oneself and for left-behind family members as evinced in the flow of money transfers and material goods (Constable, 2003; Tolentino, 1996) or the promise of enriched social influence at home, and even bearing light-skinned, blue-eyed children (Cabalquinto & Soriano, 2020). In Chapter 4, we problematise the visibilities of interracial intimacies on YouTube, which we argue demonstrate the processes of cultural whitening and offers affective aspirations for ideal intimate relationships.

World-class labour

In a nation-state like the Philippines that relies on exporting cheap and surplus human labour to sustain its economy, freelance digital work serves as an additional, alternative, and lucrative source of income. From foreign domestic labour to call centre labour (Abara & Heo, 2013) and now to digital platform labour, Filipinos aim for work opportunities that match their "distinct" traits as the top service workers of the world despite the many exploitative conditions accompanying this distinction (Soriano & Cabanes, 2020a). This constructs the notion of the Philippines as a nation systematically training and exporting service workers for the global market has been predominantly constructed as a natural global order of things that Filipinos ought to take advantage of (Fabros, 2016; Soriano & Cabañes, 2020b). An important local context to be considered here is the continuing expansion of the large informal economy and the continuing flexibilisation of work that drive the popularity of platform labour locally (Ofreneo, 2013). The many Filipino professionals and casual employees who are moving into online platform labour need to be placed side-by-side the many others who belong to the "informal economy," or those obtaining small gigs through informal networks such as food peddlers in variety stores, mobile load sellers, public transportation drivers, caregivers, domestic helpers, student-research assistants, among others, who are also eagerly jumping into opportunities for obtaining a job in online labour platforms. This is why despite critiques about poor security or the absence of long-term advancement, online labour platforms are often viewed as an attractive employment option locally in comparison to other alternatives. And yet, despite government pronouncements promoting digital labour as a crucial solution to unemployment, mechanisms for supporting workers engaged in labour platforms are absent. For labour migration, several private and public institutions have been set up to help workers aspiring to migrate overseas for jobs in terms of employment seeking, expectation-setting, salary identification, taxation, or welfare protection. For BPO-related jobs such as call centre work, foreign companies operating in the country have institutionalised recruitment and employment mechanisms (Kleibert, 2015). By contrast, aspiring workers bid for jobs in labour platforms through the help of "brokers" (Soriano, 2021). As we will discuss in Chapter 5, these brokers or intermediaries use the function of multiple social media spaces and particularly YouTube to create vignettes of success and possibility for aspiring workers. In effect, they also play the role of brokers between workers and labour platforms, while directly benefiting socially and economically from the process.

Politics and progressive governance

While the affective aspirations we have discussed in the previous sections pertain to the personal, people's behaviours and actions are equally driven by their desire for progress and development at the national level (Appadurai, 1990) that can potentially shape personal advancement. In most postcolonial societies, these visions are commonly anchored on Western ideals of democratic development and economic progress (Fujita-Rony, 2010). Colonial history serves as a backdrop and resource for the active construction of narratives of progress as marketable to its citizens. "Images of colonialism and postcolonialism are integrated in banal forms of interactions" (Vitorio, 2019, p. 106) such as everyday pronouncements, and political speeches, which are integral to how Filipinos imagine the possibilities of national progress and development. As Thurlow and Jaworski (2010) argue, "It is at the level of the interpersonal, everyday exchange of meaning where the global and the local interface are negotiated and resolved, be it through processes of cultural absorption, appropriation, recognition, acceptance, or resistance" (p. 9).

Both Spanish and American colonisers projected notions of civilisation that undermine the Philippines' own determination of development. These definitions continue to exist in the Philippines today through what some scholars call "colonial mentality," or "referring to the notion that superiority, pleasantness, or desirability are associated with cultural values, behaviours, physical appearance, and objects that are American or Western" (David & Okazaki, 2010, p. 850). Moreover, despite the liberation of the Philippines in 1945, the country remains dependent on the U.S. as reflected in foreign policies and international trade (Aguilar, 2014). At the time of former President Marcos, gargantuan and notable infrastructures were built to legitimise power attached to Western imaginations of development (Lico, 2003), while the country resorted to the exportation of cheap labour to address issues of economic crisis. It is interesting to note that transnational partnerships during previous postcolonial administration can be articulated as fantasies of progress and sustainability and modernisation, despite the increased marginalisation of ordinary Filipinos (Tadiar, 2004). Postcolonial governance is discussed by Benedicto (2013) by examining the postcoloniality of architectures particularly during the Marcos regime:

> The structure's departure from and adherence to the tenets of architectural modernism mirror the precarious position occupied by the Marcos regime as a postcolonial dictatorship. Its architecture abides by

central "international" principles such as the rejection of adornment and frivolity, but it also affects spectacular excess through scale, height, and the starkness of its contrast with the "thirdworldness" of metropolitan Manila. (p. 26)

In Chapter 6, we showcase how Filipino YouTubers mobilise affective aspirations on the nostalgia of a progressive state during the Marcos period. In particular, we show how pro-Marcos YouTubers represent aspirational tropes of "what was" (during Ferdinand Marcos' "golden years" – or an era when the Philippines was comparable to "Western states") and "what could be" (i.e. if one votes for another Marcos in the succeeding elections), while hiding the documented violent atrocities surrounding that political regime. Under these visions of "what could be" involve visualisations of national pride, regional competitiveness and equal standing, if not full independence from previous "colonial controls." Notably, political actors harness affective aspirations by advancing a political agenda that illuminates tropes of nostalgia and comparison. However, this becomes problematic especially when narratives contribute to political disinformation, while hiding its political agenda.

These four key themes – bleached skin, interracial intimacy, world-class labour, and progressive governance – are critical sites for examining the historically-situated conditions that structure the enactment of affective aspirations on YouTube. However, as we highlight in this book, they also serve as entry points to unravel paradoxical reconfigurations or enacting transformations that simultaneously negotiate and reinforce existing asymmetries in a neoliberal and neo-colonial landscape. In the next section, we present the key dimensions of analysing brokerage in the digital space.

Key Dimensions of Brokerage in the Digital Realm

We propose brokering in online spaces as a mechanism of networked individuals who capitalise on affective aspirations through personalised, creative, and branding strategies. It focuses on the utilisation of various digital technologies and online channels enacting neoliberal principles and colonialist legacies for crafting aspirations, nurturing networks, harnessing platform-specific strategies, and monetisation. In developing our conception of brokerage in a digital space, we introduce its four key dimensions, including aspirational content, discursive style, credibility-building, and platform-specific strategies. These frames are applicable for interrogating the elements and practices of YouTube content production and engagement of

Filipinos, whose lived conditions are entangled and mapped in the domains of neoliberal globalisation, colonialist conditions, and digital environments.

Firstly, digital brokering involves the production of aspirational content that is well-emplaced in postcolonial hierarchies and neoliberal imaginaries. We build on the literature on influencer culture (Abidin, 2018; Duffy, 2017), noting how YouTube content creators, as brokers, capitalise on putting together crucial information and intimate knowledge as relatable content. Certainly, contents portray certain ideals, desires, aspirations, and intimacies (Abidin, 2015; Berryman & Kavka, 2017). Despite allowing previously marginalised and colonised individuals to occupy online spaces, the content becomes recognisable or visible through the frames of hierarchical structures, which weave gendered, racialised, and classed aspirations (Gajjala, 2013; Nakamura, 2013). Further, these contents produce paradoxical consequences especially when "being seen" is appropriated for the triumphs of the invisible hands of a colonialist and economic globalisation (Gajjala, 2013).

Our approach to the analysis of content as texts is surfaced through the second frame of brokering, which is the brokers' discursive styles to highlight and amplify their key offerings. This includes the narrative styles, video components, audio templates, and other add-ons that are engaged for generating "communicative intimacy" (Abidin, 2015) between the content creator and the audience. The appeal of content lies in the ability to show the performance of relatability on screen by real people (Kavka, 2008). Moreover, engagement is fuelled by an influencer's calibrated amateurism, presenting what is raw, spontaneous and intimate (Abidin, 2017). These branding strategies (Duffy, 2017; Marwick, 2013) are then translated to networked gains, such as followers, content engagement, and profit. Further, various styles are understood to be reflecting the embodiment of neoliberal values, such as showcasing innovation, a driven spirit, and entrepreneurialism (Banet-Weiser, 2012; Duffy, 2017). Hierarchical and historically situated structures shape the utilisation of different visual, aural and textual styles that re-enact the colonial gaze (Gajjala, 2013; Sobande, 2017, 2019).

Credibility-building is the third aspect of digital brokering. The discussion foregrounds ways of enacting authenticity and legitimacy to speak in influencer cultures (Abidin, 2018; Duffy, 2017; Senft, 2008), which include visibilising "success" or "struggle" (Berryman & Kavka, 2018) and behind-the-scene intimate encounters (Abidin, 2018). We highlight credibility-building as a form of affective investment (Kavka, 2008) for the content creator to establish their legitimacy – as ordinary people – to speak and be heard by their audiences. Kavka's (2008) work is useful in this regard. By examining reality TV shows, Kavka (2008) unpacked how people engage and connect to the content because of a psychical investment or seeing performances by real people as a reality rather than a representation. The provocation of visibilised transformations and expressions of triumph that influencers know well to appropriate signal possibilities for ordinary people to negotiate the limits of existing conditions. In a sense, this becomes a potent aspect of believability, which is translated to viewing and other forms of engagement. This point is well captured in this statement: "Reality TV evokes what I call situational identification; we identify with the situation through its affective resonance; this is what it feels like to be someone sitting around a kitchen table negotiating a fellow intimate's upset/anger/sense of exclusion" (Kavka, 2008, p. 92). In this study, we define credibility-building as the act of portraying the aspirations, actions to achieve such aspirations, and "definitions of success" of ordinary Filipinos for aspiring audiences.

Lastly, platform-specific strategies are part of digital brokering. Online channels have affordances that allow individuals to produce, circulate and consume diverse content. However, in a data-driven and networked environment, platform-specific features are exploited for gaining attention, harnessing followers, and generating profit (Abidin, 2018). Individuals can be using hashtags, keywords, click buttons, and other types of prompts that link content to wider audiences. Simultaneously, algorithms render visibility, ranking, and curation of content. As argued earlier, the platform itself is a broker, offering aspirations for audiences via notions of profiteering, social status, differentiation, and domination (van Dijck, 2013). On YouTube, the neoliberal self is developed based on hits, likes, shares, and the navigation of platform affordances through the entrepreneurial spirit of its users (Duffy, 2017; Glatt, 2017). These conditions show the politics that exist in the operations and governance of these commercial platforms (Gillespie, 2010), which evidently shows who gets to be visible and invisible in digital spaces.

Brokerage as an analytical lens exposes the paradoxes that exist in online environments. Aspirational content, creative and robust discursive styles, credibility-building strategies, and platform-specific strategies are engaged for visibility and agency. However, we contend that these practices enact valuable representational and connective opportunities for Filipinos in the context of neoliberal economies and postcolonial conditions. As we will show in the succeeding chapters, the positionality of Filipino YouTubers highlights ongoing negotiation of power, as well as the contradictions, continuities and changes that relate to Philippine modernity in the context of global digital capitalism.

Revisiting Digital Brokering

Social media platforms such as YouTube have been pivotal in redefining the conduct of Philippine contemporary society. In this Chapter, we presented brokerage as a critical lens for analysing online content production and performativity on YouTube, including how this produces *affective aspirations* and *paradoxical reconfigurations*. First, we built our discussion on the growing literature on influencer culture and platform studies, as well as on the sociological literature on brokerage. We argued that brokers enact the role of influencers through branding strategies and mastery of the technological affordances of an online platform. We also argued that the platform itself is a broker, facilitating the flows and control of information. But at the centre of articulating brokerage is considering how economic, political, cultural, and socio-technological forces inform the production of affective aspirations and paradoxical outcomes. Significantly, the book pays special attention to how socio-historical frames can be used to re-define notions of brokering in the digital era. Inasmuch as we posit that mediated practices are embedded in a postcolonial condition, it is also equally situated in a neoliberal system. By taking the Philippines as a case study for investigation, we highlight how the brokerage of aspirations in various aspects – bleached skin, interracial intimacy, world-class labour, and progressive governance – manifests how ordinary people navigate their dreamwork of progress, mobility, and stability in a digital, neoliberal, and neo-colonial terrain. Crucial in the process is how aspirations, practices, and networking allow them to relive and negotiate the legacies of colonial history and the conditions of digital capitalism. To unpack this, we outlined and explained the four key dimensions of brokerage in a digital hub – aspirational content, discursive style, credibility-building, and platform-specific strategies. These elements are showcased and discussed across the succeeding case study chapters.

References

Abara, A. C. & Heo, Y. (2013). Resilience and recovery: The Philippine IT-BPO industry during the global crisis. *International Area Studies Review, 16*(2), 160–183.

Abidin, C. (2015). Communicative ❤ intimacies: Influencers and perceived interconnectedness. *Ada: A Journal of Gender, New Media, and Technology, 8*, 1–16. doi:10.7264/N3MW2FFG

Abidin, C. (2017). #familygoals: Family influencers, calibrated amateurism, and justifying young digital labour. *Social Media + Society.* doi:10.1177/2056305117707191

Abidin, C. (2018). *Internet celebrity: Understanding fame online* (First ed.). Emerald Publishing Limited.

Aguilar, F. J. (2014). *Migration revolution: Philippine nationhood and class relations in a globalized age*. Ateneo de Manila University Press.

Amoore, L. (2011). Data derivatives: On the emergence of a security risk calculus for our times. *Theory, Culture & Society, 28*(6), 24–43.

Appadurai, A. (1990). Disjunction and difference in a global cultural economy. In J. Featherstone (Ed.), *Global culture: Nationalism, globalism and modernity* (Vol. 295–310). Sage.

Appadurai, A. (1996). *Modernity at large: Cultural dimensions of globalization*. University of Minnesota Press.

Aquino, K. (2018). *Racism and resistance among the Filipino diaspora: Everyday anti-racism in Australia*. Routledge.

Arnado, J. M. (2019). Cultural whitening, mobility and differentiation: Lived experiences of Filipina wives to white men. *Journal of Ethnic and Migration Studies*, 1–17. doi:10.1080/1369 183X.2019.1696668

Arthurs, J., Drakopoulou, S., & Gandini, A. (2018). Researching YouTube. *Convergence, 24*(1), 3–15. doi:10.1177/1354856517737222

Baldo-Cubelo, J. T. (2015). The embodiment of the New Woman: Advertisements' mobilization of women's bodies through co-optation of feminist ideologies. *Plaridel: A Philippine Journal of Communication, Media, and Society, 12*(1), 42–65.

Banet-Weiser, S. (2012). *Authentic TM: The politics and ambivalence in a brand culture*. New York University Press.

Benedicto, B. (2013). Queer space in the ruins of dictatorship architecture. *Social Text, 31*(4 [117]), 25–47. doi:10.1215/01642472-2348977

Berryman, R., & Kavka, M. (2017). 'I guess a lot of people see me as a big sister or a friend': The role of intimacy in the celebrification of beauty vloggers. *Journal of Gender Studies, 26*(3), 307–320. doi:10.1080/09589236.2017.1288611

Berryman, R., & Kavka, M. (2018). Crying on YouTube: Vlogs, self-exposure and the productivity of negative affect. *Convergence, 24*(1), 85–98.

Bhabha, H. (1994). *The location of culture*. Routledge.

Bishop, S. (2019). Managing visibility on YouTube through algorithmic gossip. *New Media & Society, 21*(11/12), 2589–2606.

Bourdieu, P. (1984). *A social critique of the judgement of taste*. Routledge.

Bucher, T. (2012). Want to be on the top? Algorithmic power and the threat of invisibility on Facebook. *New Media & Society, 14*(7), 1164–1180. doi:10.1177/1461444812440159

Bucher, T. (2016). Neither black nor box: Ways of knowing algorithms. In S. Kubitschko & A. Kaun (Eds.), *Innovative methods in media and communication research* (pp. 81–98). Springer International Publishing.

Bucher T. (2018) *If...then: Algorithmic power and politics*. Oxford University Press.

Burgess, J., & Green, J. (2009). *YouTube: Online video and participatory culture*. Polity.

Burt, R. (2007). *Brokerage and closure: An introduction to social capital*. Oxford University Press.

Cabalquinto, E. C., & Soriano, C. R. R. (2020). 'Hey, I like ur videos. Super relate!' Locating sisterhood in a postcolonial intimate public on YouTube. *Information, Communication & Society, 23*(6), 892–907. doi:10.1080/1369118X.2020.1751864

Camba, A. (2012). Religion, disaster, and colonial power in the Spanish Philippines in the sixteenth to seventeenth centuries. *Journal for the Study of Religion, Nature and Culture, 6*(2), 215–231. doi:10.1558/jsrnc.v6i2.215

Casilli, A. (2017). Digital labour studies go global: Towards a digital decolonial turn. *International Journal of Communication 11*, 3934–3954.

Cheney-Lippold, J. (2017). *We are data: Algorithms and the making of our digital selves.* New York University Press.

Cohn, J. (2019). *The burden of choice: Recommendations, subversion, and algorithmic culture.* Rutgers University Press.

Constable, N. (2003). *Romance on a global stage: Pen pals, virtual ethnography, and 'mail-order' marriages.* University of California Press.

Constable, N. (2006). Brides, maids, and prostitutes: Reflections on the study of 'trafficked' women. *Portal: Journal of Multidisciplinary International Studies, 3*(2), 1–25. doi:10.5130/portal.v3i2.164

Constable, N. (2009). The commodification of intimacy: Marriage, sex, and reproductive labour. *Annual Review of Anthropology, 38,* 49. doi:10.1146/annurev.anthro.37.081407.085133

Constable, N. (2012). Correspondence marriages, imagined virtual communities, and countererotics on the Internet. In P. Mankekar & L. Schein (Eds.), *Media, erotics, and transnational Asia* (pp. 111–138). Duke University Press.

David, E. J. R. (2013). *Brown skin, white minds: Filipino-American postcolonial psychology.* Information Age Pub. Inc.

David, E. J. R., & Okazaki, S. (2006). The Colonial Mentality Scale (CMS) for Filipino Americans: Scale construction and psychological implications. *Journal of Counselling Psychology, 53*(2), 241–252. doi:10.1037/0022-0167.53.2.241

Duffy, B. E. (2017). *(Not) getting paid to do what you love: Gender, social media, and aspirational work.* Yale University Press.

Fabros, A. (2016). *Outsourceable selves: An ethnography of call center work in a global economy of signs and selves.* Ateneo de Manila University Press.

Fisher, M. & Taub, A. (2019). How YouTube radicalized Brazil. Retrieved 12 July 2020, https://www.nytimes.com/2019/08/11/world/americas/YouTube-brazil.html

Fujita-Rony, D. (2010). History through a postcolonial lens: Reframing Philippine Seattle. *The Pacific Northwest Quarterly, 102*(1), 3–13.

Gajjala, R. (2013). *Cyberculture and the subaltern: Weavings of the virtual and real.* Lexington Books.

Gillespie, T. (2010). The politics of 'platforms.' *New Media & Society, 12*(3), 347–364. doi:10.1177/1461444809342738

Gillespie, T. (2014). The relevance of algorithms. In T. Gillespie, P. Boczkowski, & K. Foot (Eds.), *Media technologies: Essays on communication, materiality, and society* (pp. 167–193). MIT Press.

Gillespie, T. (2016). #trendingistrending: When algorithms become culture. In R. Seyfort & J. Roberge (Eds.), *Algorithmic cultures: Essays on meaning, performance, and new technologies* (pp. 52–75). Routledge.

Glatt, Z. (2017). *The commodification of YouTube Vloggers.* (Masters), University of London, Retrieved 18 June 2020, https://zoeglatt.com/wp-content/uploads/2020/05/Glatt-2017-The-Commodification-of-YouTube-Vloggers.pdf

Glenn, E. N. (2008). Yearning for lightness: Transnational circuits in the marketing and consumption of skin lighteners. *Gender & Society, 22*(3), 281–302. doi:10.1177/0891243208316089

Glenn, E. N. (2009). Consuming lightness: Segmented markets and global capital in the skin-whitening trade. In G. E. N. (Ed.), *Shades of difference: Why skin color matters* (pp. 166–187). Stanford University Press.

Gonzalez, V. V., & Rodriguez, R. M. (2003). Filipina.com: Wives, workers and whores on the cyber frontier. In R. Lee & S.-l. C. Wong (Eds.), *Asian America.Net: Ethnicity, nationalism, and cyberspace* (pp. 215–234). Routledge.

Guevarra, A. R. (2010). *Marketing dreams, manufacturing heroes: The transnational labour brokering of Filipino workers.* Rutgers University Press.

Hou, M. (2019). Social media celebrity and the institutionalization of YouTube. *Convergence*, 25(3), 534–553. doi:10.1177/1354856517750368

Hutchinson, J. (2017). *Cultural intermediaries: Audience participation in media organisations.* Palgrave Macmillan.

IBON Foundation (2017). Contractualization prevails. *Facts and Figures* 7(7). Retrieved 16 May 2020, https://www.ibon.org/contractualization-prevails-ibon-facts-figures-excerpt/

Information Technology and E-Commerce Council (2003). ITECC Strategic Road Map: Linking Government Services for OFWs. Retrieved 12 June 2019, www.ncc.gov.ph/files/strat_road-mapReport.pdf

Jamerson, T. W. (2019). Race, markets, and digital technologies: Historical and conceptual frameworks. In G. D. Johnson, K. D. Thomas, A. K. Harrison, & S. A. Grier (Eds.), *Race in the marketplace.* (pp. 39–54). Palgrave Macmillan.

Jamieson, K.L. (2018). *Cyberwar: How Russian hackers and trolls helped elect a president; What we don't, can't, and do know.* Oxford University Press.

Jenkins, H. (2006). *Convergence culture: Where old and new media collide.* New York University Press.

Jensen, S. (2018). Epilogue: Brokers–pawns, disrupters, assemblers? *Ethnos, 83*(5), 888–891. doi:10.1080/00141844.2017.1362455

Kapunan, R. (1991). Labour-only contractors: New generation of *"cabos." Philippine Law Journal* 65(5), 326.

Kaufman, R. (1974). The patron-client concept and macro-politics: Prospects and problems. *Comparative Studies in Society and History, 16*(3), 284–308. doi:10.1017/S0010417500012457

Kavka, M. (2008). *Reality television, affect and intimacy: Reality matters.* Palgrave Macmillan.

Kern, A., & Müller-Böker, U. (2015). The middle space of migration: A case study on brokerage and recruitment agencies in Nepal. *Geoforum, 65,* 158–169. doi:10.1016/j.geoforum.2015.07.024

Khamis, S., Ang, L., & Welling, R. (2017). Self-branding, 'micro-celebrity' and the rise of social media influencers. *Celebrity Studies, 8*(2), 191–208.

Kleibert, J. M. (2015). Services-led economic development: Comparing the emergence of the offshore service sector in India and the Philippines. In Lambregts, B., Beerepoot, N., Kloosterman, R.C. (Eds.), *The Local impact of globalization in South and Southeast Asia: Offshore business process outsourcing in services industries* (pp. 29–45). Routledge.

Laforteza, E. (2016). *The somatechnics of whiteness and race: Colonialism and Mestiza privilege.* Taylor and Francis Routledge.

Lasco, G., & Hardon, A. P. (2020). Keeping up with the times: Skin-lightening practices among young men in the Philippines. *Culture, Health and Sexuality, 22*(7), 838–853. doi:10.1080/13691058.2019.1671495

Lash, S. (2007). Power after hegemony: Cultural studies in mutation? *Theory, Culture & Society, 24*(3), 55–78.

Lewis, R. (2018). *Alternative influence: Broadcasting the reactionary right on YouTube.* Retrieved 6 April 2020, https://datasociety.net/wpcontent/uploads/2018/09/DS_Alternative_Influence.pdf

Lewis, R. (2020). 'This is what the news won't show you': YouTube creators and the reactionary politics of micro-celebrity. *Television & New Media, 21*(2), 201–217. doi:10.1177/1527476419879919

Lico, G. (2003). *Edifice complex: Power, myth, and Marcos state architecture.* Ateneo de Manila University Press.

Lindquist, J., Xiang, B., & Yeoh, B. S. A. (2012). Opening the black box of migration: Brokers, the organization of transnational mobility and the changing political economy in Asia. *Pacific Affairs, 85*(1), 7–19. doi:10.5509/20128517

Lobato, R. (2016). The cultural logic of digital intermediaries: YouTube multichannel networks. *Convergence 22*(4), 348–360

Loomba, A. (1998). *Colonialism/postcolonialism. The new critical idiom.* Routledge.

Malefyt, T. & Morais, R. (2012). *Advertising and anthropology: Ethnographic practice and cultural perspectives.* Berg.

Marwick, A. E. (2013). *Status update: Celebrity, publicity, and branding in the social media age.* Yale University Press.

Marwick, A. E. (2015). Instafame: Luxury selfies in the attention economy. *Public Culture, 27*(1), 137–160. doi:10.1215/08992363-2798379

McKay, D., & Perez, P. (2019). Citizen aid, social media and brokerage after disaster. *Third World Quarterly, 40*(10), 1903–1920. doi:10.1080/01436597.2019.1634470

Mendoza, R. L. (2014). The skin whitening industry in the Philippines. *Journal of Public Health Policy, 35*(2), 219–238.

Nakamura, L. (2013). *Cybertypes: Race, ethnicity, and identity on the Internet.* Routledge.

Nakamura, L., & Chow-White, P. (2012). *Race after the Internet.* Routledge.

Negus, K. (2002). The work of cultural intermediaries and the enduring distance between production and consumption. *Cultural Studies*, 16:4, 501–515. doi:10.1080/09502380210139089

Nowak, T.C. & Snyder, K.A. (1974). Clientelist politics in the Philippines: Integration or instability? *American Political Science Review 68*(3): 1147–1170.

Ofreneo, R.E. (2013). Precarious Philippines: Expanding informal sector, 'flexibilizing' labour market. *American Behavioral Scientist, 57*(4), 420–443. doi:10.1177/0002764212466237

Ong, J. C., & Cabañes, J. V. (2018). *The architects of networked disinformation: Behind the scenes of troll accounts and fake news production in the Philippines.* Retrieved 10 June 2020, http://newtontechfordev.com/wp-content/uploads/2018/02/ARCHITECTS-OF-NETWORKED-DISINFORMATION-FULL-REPORT.pdf

Ong, J. C., & Combinido, P. (2018). Local aid workers in the digital humanitarian project: Between 'second class citizens' and 'entrepreneurial survivors.' *Critical Asian Studies, 50*(1), 86–102. doi:10.1080/14672715.2017.1401937

Ong, J. C., Tapsell, R., & Curato, N. (2019). *Tracking digital disinformation in the 2019 Philippine midterm election.* Retrieved 12 June 2020, https://www.newmandala.org/wp-content/uploads/2019/08/Digital-Disinformation-2019-Midterms.pdf

Oreglia, E. & Ling, R. (2018). Popular digital imagination: Grassroot conceptualization of the mobile phone in the Global South. *Journal of Communication, 68.* doi:10.1093/joc/jqy01

Pasquale, F. (2015). *The black box society: The secret algorithms that control money and information.* Harvard University Press.

Pertierra, R. (2020). Anthropology and the AlDub nation, entertainment as politics and politics as entertainment. *Philippine Studies: Historical & Ethnographic Viewpoints, 64,* 289–300.

Rafael, V. (2000). *White love and other events in Filipino history.* Duke University Press.

Reviglio, U., & Agosti, C. (2020). Thinking outside the black-box: The case for 'algorithmic sovereignty' in social media. *Social Media + Society,* 1–12. doi:10.1177/2056305120915613

Rieder, B., Matamoros-Fernández, A. & Coromina, Ò. (2018). From ranking algorithms to 'ranking cultures': Investigating the modulation of visibility in YouTube search results. *Convergence 24*(1): 50–68.

Robinson, K. (1996). Of mail-order brides and 'boys' own' tales: Representations of Asian-Australian marriages. *Feminist Review*(52), 53–68. doi:10.1057/fr.1996.7

Rodriguez, R. M. (2010). *Migrants for export: How the Philippine state brokers to the world.* The University of Minnesota Press.

Rondilla, J. (2009). Filipinos and the color complex, ideal Asian beauty. In G. E. N. (Ed.), *Shades of difference: Why skin color matters* (pp. 63– 80). Stanford University Press.

Rondilla, J. L. (2012). *Colonial faces: Beauty and skin color hierarchy in the Philippines and the U.S.* (Doctoral dissertation), University of California, Berkeley, CA.

Rondilla, J. L., & Spickard, P. (2007). *Is lighter better?: Skin-tone discrimination among Asian Americans.* Rowman & Littlefield Publishers.

Saban, L. I. (2015). Entrepreneurial brokers in disaster response network in typhoon Haiyan in the Philippines. *Public Management Review, 17*(10), 1496–1517. doi:10.1080/14719037.2014.943271

San Juan, E. (2009). Overseas Filipino workers: The making of an Asian-Pacific diaspora. *The Global South, 3*(2), 99–129.

Saroca, C. (2002). *Hearing the voices of Filipino women: Violence, media representation and contested realities.* (Doctor of Philosophy), University of Newcastle, Newcastle.

Saroca, C. (2007). Filipino women, migration and violence in Australia: Lived reality and media image. *Kasarinlan: Philippine Journal of Third World Studies, 21*(1), 75–110.

Senft, T. M. (2008). *Camgirls: Celebrity & community in the age of social networks.* Peter Lang.

Shome, R. & Hegde, R.S. (2002). Postcolonial approaches to communication: Charting the terrain, engaging the intersections. *Communication Theory, 12*: 249–270. doi:10.1111/j.1468-2885.2002.tb00269.x

Shrestha, T., & Yeoh, B. S. A. (2018). Introduction: Practices of brokerage and the making of migration infrastructure in Asia. *Pacific Affairs, 91*(4), 663–672.

Silarde, V.Q. (2020). Historical roots and prospects of ending precarious employment in the Philippines. *Labour and Society 23*(4), 461–484.

Sobande, F. (2017). Watching me watching you: Black women in Britain on YouTube. *European Journal of Cultural Studies, 20*(6), 655–671. doi:10.1177/1367549417733001

Sobande, F. (2019). Constructing and Critiquing Interracial Couples on YouTube. In G. D. Johnson, K. D. Thomas, A. K. Harrison, & S. A. Grier (Eds.), *Race in the Marketplace.* (pp. 73–85). Palgrave Macmillan.

Soriano, C. R. R. (2021). Digital labour in the Philippines: Emerging forms of brokerage. *Media International Australia.* doi:10.1177/1329878X21993114

Soriano, C.R. & Cabañes, J.V. (2020a). Entrepreneurial solidarities: Social media collectives and Filipino digital platform workers. *Social Media + Society.* doi:10.1177/2056305120926484

Soriano, C.R. & Cabañes, J.V. (2020b). Between 'world class work' and 'proletarianized labour': Digital labour imaginaries in the Global South. In Polson E, Schofield-Clarke, L., & Gajjala, R. (Eds.) *The Routledge companion to media and class* (pp. 213–226). Routledge.

Stovel, K., & Shaw, L. (2012). Brokerage. *Annual Review of Sociology, 38*(1), 139–158. doi:10.1146/annurev-soc-081309-150054

Tadiar, N. X. M. (2004). *Fantasy production: Sexual economies and other Philippine consequences for the New World Order*: Hong Kong University Press.

Thurlow, C., & Jaworski, A. (2017). The discursive production and maintenance of class privilege: Permeable geographies, slippery rhetorics. *Discourse & Society, 28*(5), 535–558. doi:10.1177/0957926517713778

Tolentino, R. B. (1996). Bodies, letters, catalogs: Filipinas in transnational space. *Social Text*(48), 49–76. doi:10.2307/466786

Utz, S. & Wolfers, L. (2020). How-to videos on YouTube: The role of the instructor. *Information, Communication & Society,* 1–16. doi:10.1080/1369118X.2020.1804984

van Dijck, J. (2013). *The culture of connectivity: A critical history of social media.* Oxford University Press.

Vitorio, R. (2019). Postcolonial performativity in the Philippine heritage tourism industry. In A. Mietzner & A. Storch (Eds.), *Language and tourism in postcolonial settings* (pp. 106–129). Channel View Publications.

3   Self

Abstract

This chapter underscores the brokering of ideal and dominant discourses of femininity among Filipina YouTubers. Applying the concept of "cultural whitening" and a postfeminist critique, it analyses the diverse affective, creative and ambivalent tactics performed by YouTubers. It first situates the examination of YouTubers' imaginaries of standard femininities within the socio-historical context of gendered, racial and classed hierarchy in the Philippines. On top of magnifying the tropes, tactics, and tools articulated by YouTubers in legitimising bodily care routines and showcasing transformations, it also captures the platform-specific mechanisms and marketised logics in the brokerage of beauty ideals. The chapter concludes by underlining the paradoxical outcomes of gendered, racialised, and classed visibility, performativity, and engagement on YouTube.

Keywords: underarm whitening, postfeminism, mestisa, cultural whitening, colourism, entrepreneurial femininity

Diverse modes of representing standards of feminine beauty continue to dominate Philippine media. More specifically, creative advertisements designed, produced and distributed by business enterprises and creative agencies encourage Filipino women to achieve a recognisable beauty through the use of a range of skin lightening products. Belo, Loreal, Nivea, and Olay are some of the big brands that endorse the realisation and positive benefits of fairer skin among women and men. These products are promoted and distributed by the beauty and media industry across multiple channels, including online and offline. As creative, compelling and star-studded campaigns communicate and sell whitening products online, profit flows into the processes and systems of the media culture industry, including but not limited to health, fashion, and so forth. In a sense, the production and circulation of celebrity-driven endorsements align with our conception of brokerage, which emphasises the collection, curation and circulation

Soriano, Cheryll Ruth and Earvin Charles Cabalquinto: *Philippine Digital Cultures: Brokerage Dynamics on YouTube*. Amsterdam: Amsterdam University Press, 2022
DOI: 10.5117/9789463722445_CH03

of aspirational contents for certain individuals, groups and markets in diverse media channels. However, brokerage in this context contains the marketability of standard beauty within the imaginaries of individuals in the dominant creative and media industry.

The brokerage of beauty standards – skin practices – is governed by the beauty and media industry and noticeable in the Philippine context. In 2019 alone, two particular ads created a buzz in Philippine history of skin whitening products. Skin White released their "Dark or White, you are beautiful" campaign. It featured twins Marianne and Martha Bibal, one with a white face and the other with a brown skin tone. In the same year, Glutamax also released the campaign "Your fair treatment," highlighting that "3 in 5 Filipinos believe that people with fairer skin receive better treatment than others." This particular campaign of Glutamax succeeded their popular skin whitening ad in 2009. In that campaign, Filipina celebrity Jinky Oda was presented with a transformation, from being ebony to becoming ivory. These campaigns primarily highlighted the duty of care of women to their body in order to be beautiful and successful in life. Situated in the centre of the advertising texts was the endorsement of beauty products and the best practices to use these consumer goods to achieve the best results. The celebrity or endorser brokers the possibilities of jaw-dropping transformations through the persistent consumption of the endorsed beauty items.

However, advertisements can also draw flak among discerning individuals who proactively identify and critique unfair and insensitive media messages. Filipinos criticise the advertisements across traditional media channels and online media for the stereotypical and discriminatory narratives on a Filipina's beauty standards. An example of this is the recent backlash on the 2021 "Pandemic Effect" ad of the Belo Medical Group. The video shows a woman sitting on the couch and watching a slew of bad news. With the constant exposure to media reports, the woman's appearance changes, including growing hair all over her body and gaining weight. The ad ends with the tagline, "Tough times call for beautiful measures," and also paired with the call-to-action that encourages viewers to book a consultation with the medical group. Celebrities and several online stars have been part of the campaign, showing their own "pandemic effect" stories and validating the need for self-care in tough times. Although not particularly pertaining to skin whitening and focusing more broadly on the women's full body, the ad have received criticisms from Filipinos because of "body shaming" women in the middle of the pandemic. Online channels have been filled with commentaries from individuals, appropriating the hashtag #pandemiceffect into expressions of frustrations and disapproval. The video has eventually been taken down.

To date, paradoxically, the Philippine media remains filled by a range of stories, creative texts, and performances that still privilege fair-skinned and often mixed-raced figures. Imaginaries of beauty standards – mestiso/mestisa looking – are deeply ingrained in media systems and texts. Celebrities as brokers in the creative, media, and beauty industries are accomplices in the articulation of legitimised beauty. But despite their strong positionality in "selling" what's acceptable and favoured, their performances are also deeply tied to their markets, either appealing or "hurting" them. In noting this landscape and also tensions between the media personalities and audiences, we ask, what has changed after all despite the expressions of revulsion for discriminatory media contents on beauty standards? Further, has the democratising potential of the Internet and social media paved the way for a reconfiguration of the expressions of beauty among Filipino women?

This chapter presents the modes of brokering an ideal feminine beauty among Filipina YouTubers. It specifically shows how ten Filipina YouTubers communicate, promote and sell feminine beauty by visibilising and curating the whitening and smoothening of their underarm through bodily care routines. We chose this topic for two particular reasons. First, we are interested in unpacking the brokering of an ideal Filipina identity in the Philippines through representations of feminised beauty through "whitening" practices on YouTube. As presented at the outset, whitening the underarm is one of the key themes that populate the promotion of beauty products in the Philippines. Despite not being visible all the most, celebrities in advertisements, whose narratives are influenced and controlled by the beauty and media industry, position a dark underarm as a problem – the cause of embarrassment or a low self-esteem among Filipinas (Baldo-Cubelo, 2015). However, in this chapter, the focus is how micro-celebrities (Abidin, 2018), who act as brokers by collating, curating and distributing affective contents built on their personal experiences and ordinary expertise. Second, as the topic on skin whitening was too broad, we used Google Trends to narrow down the topic. We initially used the word *pampaputi* (whitening). Through this process, *pampaputi ng kili-kili* or underarm whitening appeared to be the most popularly searched keywords on YouTube during the conduct of the study in August 2020. This revelation showed that underarm whitening has likely become a priority for individuals who search such content and potentially aspire for such skin transformation. We then accessed YouTube and collected videos. Based on collecting 40 videos through algorithmic recommendations on YouTube, we came up with ten videos, which we closely and critically analysed.

We approach our discussion by positioning Filipina YouTubers as curators and brokers of beauty regimens that reflect cultural whitening. Deploying the conception of cultural whitening (Arnado 2019), we investigate how Filipina YouTubers use YouTube as a platform to recommend ways of achieving a fairer skin, a mechanism to broker beauty standards and facilitate access to a range of capitals. As argued by many Filipino scholars, the act of desiring and embodying a fairer skin deeply connects to the Philippine's colonial history (Rafael, 2000; Rondilla, 2012). A fair or white skin tone is ascribed with positive meanings as opposed to a dark shade. And skin colour typically becomes a currency especially in one's position and advancement in Philippine society, which is reflected in the number of mestiso/as thriving in Philippine cultural, economic and political scene (Rafael 2000; Rondilla, 2009). As such, wanting to have a fair or whiter skin is symptomatic of reinforcing hegemonic racial, gendered and classed discourses.

We also apply a postfeminist critique (Banet-Weiser, 2012) of visibilising a "bleached" Filipina beauty on YouTube. This consideration allows us to interrogate the brokering and framing of standardised feminised identities that often involve the policing and control of one's bodies through a range of prescriptive individualised and entrepreneurial practices. More importantly, we highlight the role personalised representations on YouTube indicate the appropriation and negotiation of racial hierarchy in feminine subjectivity in postcolonial and neoliberal Philippines. Notably, the process of representations by broadcasting practices of consuming whitening is shaped by ambivalence (Banet-Weiser, 2012; Glatt and Banet-Weiser, 2021) activating transformations that generate views, likes and shares, yet reinforcing dominant discourse on beauty and femininity. The loop of beauty standards is also amplified by the algorithmic modality of YouTube, often circulating contents based on currently viewed videos. As such, the presentation of whitening practices essentially reinforce the dominant ideals of flawlessness and femininity.

In the next sections, we map the racial hierarchy in Philippine society, allowing us to dissect the nuances of brokering feminine subjectivity through cultural whitening among Filipina YouTubers. We also connect the discussion to the allure of having a fair skin as a form of racial capital, which contributes to the legitimacy of YouTuber's brokering strategies. We then interrogate the discussion of dominant discourse of femininity through a postfeminist lens, emphasising the production of self-motivated, entrepreneurial and even colour conscious women. Nevertheless, these sections will serve as foundations in foregrounding issues on notions of beauty, modernity and success among Filipina in a digital age.

The Shades of Racial Hierarchy

In this chapter, we highlight how the colonial history of the Philippines and how this aspect of Philippine society is considered to investigate the brokering of beauty standards of Filipino women through underarm whitening on YouTube. We consider underarm whitening through different, affordable and often DIY products as an indication of mediating notions of invested, commodified, networked, and ambivalent femininity.

The desire for a fair skin in Philippine society is deeply tied to colonial legacy. According to Philippine scholar Vicente Rafael (2000), the racial hierarchy that was implemented during the Spanish colonial rule contributed to the valorisation of whiteness. Apart from the stark contrast attached to having a white and dark skin, the high regard for whiteness was evident in the way mixed-race individuals dominated the many aspects of Philippine society (Rafael, 2000). Mestisos/as were placed in privileged status, making others celebrate or envy them. In contemporary Philippine society, mixed-race figures are seen as celebrities, newscasters, politicians, executives, and so forth (Rondilla, 2012; Rondilla, 2009) as well as in advertisements that promote standards of Filipina beauty (Baldo-Cubelo, 2015; Rondilla, 2009). Moreover, the marginalisation of dark-skinned individuals compels a set of principles and practices that essentially celebrate skin whitening (Rondilla & Spickard, 2007; David, 2013).

Racial hierarchy is reflected in colourism or stratification based on skin tone. Scholars have argued that marginalisation based on one's complexion is primarily part of colonialism and slavery. Historically, and in a global context, Europeans and white Americans created the racial hierarchy to legitimise the control and treatment of colonised and enslaved non-white people (Hunter, 2005). During the British colonial rule in Southern Africa, dark-skinned people were treated as slaves. Their skin tone was also associated with ugliness, violence, and contamination (Glenn, 2009; Dixon & Telles, 2017). In contrast, mixed-race individuals had relatively better access to a diverse range of benefits. For instance, the Mulattos or offspring of white men and slave women were accorded with better treatment such as being assigned to work as servants and artisans (Glenn, 2009). In Latin America, mestisaje or the intermixture of Spanish colonists and indigenous peoples along with unacknowledged and mixture with African slaves were privileged to access education, civility and an urban lifestyle (Glenn, 2009; Hunter, 2005). Across Asia, skin colour stratification was shown in Japanese women wearing a white make-up, the distinction between lighter-skinned Aryans and Dravidians of the South in India, and the Philippines' mestiso/a and indigenous population (Glenn, 2009). More recently across Asia, Pan-Asian beauty or individuals with both Asian and Western characteristics tend to dominate beauty standards (Yip, Ainsworth, & Hugh, 2019), which is presented in various mass media outlets (Rondilla, 2009). Nonetheless, as a result of an entrenched division based on skin, dark-skinned individuals are constrained in accessing a range of social welfare services and opportunities compared to fair-skinned people or individuals with Anglo features (Hunter, 2007, 2011).

Agency is enacted through a spectrum of strategies among those individuals who are often discriminated because of their skin colour. The use of a range of skin-lightening products (Glenn, 2009; Rondilla & Spickard, 2007), undergoing for cosmetic surgery (Glenn, 2009), and seeking a light-skinned marital partner (Glenn, 2009) are some of the practices that dark-skinned people tend to act upon. Scholars argue that this set of practices to become "white" reflects internalised racism (Glenn, 2009; Hunter, 2011; Hunter, 2007; Harris, 2009), as a response to negotiating colourism, a sense of belongingness, and intergenerational mobility (Glenn, 2009; Harris, 2009). For instance, Hall (1995) coins the term "bleaching syndrome" to articulate how African American women tend to obsess over skin-whitening products to partake in and embody the white and dominant culture, despite the damaging and irreversible effects of bleaching the body. Complementing the articulation of Hall (1995), Hunter (2011) coined the term "racial capital," which drives the internalised skin whitening practices. Hunter (2011) defines racial capital as "a resource drawing for the body that can be related to skin tone, facial features, body shape, etc." (p. 145). For her, having a light skin tone "can be transformed into social capital (social networks), symbolic capital (esteem or status) or even economic capital (high paying job or promotion)" (Hunter, 2011, p. 145). Here, a light-skinned individual can access a range of benefits such as a high-paying job, a husband, an urban lifestyle, and so forth. Through colourism, as Hunter (2005) reiterates, individuals are placed in a "beauty queue" or the rank ordering of women where the lightest gets the most perks and rewards and the darkest women get the least.

The discriminating effects on dark-skinned individuals by colourism have contributed to the million-dollar industry of skin whitening across the globe, which is part of the global beauty commerce that promotes fashion, dieting, and correcting cosmetic surgery (Jha, 2016). For instance, in 2018, the global skin lightening industry was worth USD$8.3 billion (Ng & Lachica, 2020). In a report by the Global Industry Analysts (2020), the skin market for skin lighteners is estimated at USD$8.6 billion in 2020 and is projected to reach USD$12.3 billion by 2027. Notably, promoting and bolstering the industry is strategic advertising. Advertisements typically tell consumers that having a dark skin is a problem and which solution lies at the use of whitening products (Rondilla, 2009; Hall, 1995; Glenn, 2008). Strategies in selling skin whitening products include positioning the "prestige of the product" and the intrinsic nature of a white skin. For the former, narratives showcase the transformation of the dark skin or spots and the display of doctors in their laboratory outfits (Glenn, 2009). For the latter, there is an emphasis on the use of plant extracts while the light-skinned woman is against natural scenery (Glenn, 2009). In the Philippines, light-skinned women who promote the use of skin-lightening and personal care products are framed as "new women" – flawless, wanted, and successful professionals (Baldo-Cubelo, 2015). Nevertheless, the solution offered to anxiety, discomfort and frustrations over hyperpigmentation is possible through consuming whitening products.

In the Philippine context, the features of a mixed-race individual are aspired for by Filipinos. In the first instance, a light skin is desired by Filipinos because of its association to indoor and white-collar job, success, and membership in the upper class (Rondilla, 2009). More recently, the influx of media messages as well as products from East Asia reconfigures the standards of Filipina beauty. There is an appreciation of an ideal Asian beauty through the looks of Japanese and Korean celebrities (Glenn, 2009). Media contents have also involved East Asian beauties in promoting products and services. Rondilla (2009) uses the term "relatable ideal" wherein Filipinos can relate to an Asian face and alignment with understanding of whiteness. However, despite the visibility and participation of mixed-raced Asians in various media-related projects, scholars have argued that this hybridity is paradoxical in nature (Jha, 2016; Rondilla & Spickard, 2007; Rondilla, 2009). While it presents opportunities for an inclusive, new and multicultural mode of representations, it also tends to valorise those defined primarily by Anglo features (Hunter, 2011). Further, as Filipinos tend to aspire for a Pan-Asian beauty, consumerist practices demonstrate the valorisation of having a whiter skin (Yip, Ainsworth, & Hugh, 2019). In Asia, the promotion of Pan-Asian beauty, which highlights the mixture of Asian and Western features, tends to discriminate against those individuals with a darker skin colour (mostly from Southeast Asia) and essentially promote homogenisation (Yip, Ainsworth, & Hugh 2019).

We deploy the concept of cultural whitening (Arnado, 2019) in discussing the brokering of ideal femininity in Philippine context. Here, the habits of consuming whitening products signal the embodied and objectified states of cultural whitening (Arnado, 2019). Through such processes, an individual desires to achieve a lighter or white skin, which becomes a capital for enabling connections, symbolic capital, and social mobility (Hunter, 2011). It is through these perceived outcomes of using whitening products that scholars have argued the success of the skin whitening industry in the Philippines. According to Mendoza (2014), the boom of skin lightening products in the Philippines can be attributed to postcolonial and internalised racism, which advertisers and sellers invest on in promoting and selling consumer goods. Particularly, by surveying 147 Filipinos in Manila and Quezon City, Mendoza (2014) observed that there is a favourable response to purchasing and using skin whitening products. Several buyers are convinced of the potential outcomes of consistently using a skin whitening product. However, Mendoza (2014) also highlighted the dangers of toxic and illegal whiteners that remain accessible via small retailers and informal sellers. For him, there is a need for the Food and Drug Authority (FDA) in the Philippines to address such issue. This is most important especially with the presence of high levels of harmful chemicals in whitening products, such as hydroquinone, cortico-steroids, and mercury. Further, he noticed that there exists an asymmetrical presentation of information among buyers and users of skin whitening goods. He suggests that right and useful information should be highlighted in packaging. For instance, products should convey the benefits of using products for skin protection from UV rays instead of simply reiterating the whitening effect (Rondilla, 2009). Notably, while skin whitening products remain a valuable product in Philippine society, in countries like Uganda, Kenya, South Africa, and Gambia, skin whitening products were already banned (Hunter, 2011). Ultimately, the widespread and public promotion and use of skin whitening products succeeds in the Philippines because of the promise of skin transformation and ascribed positive meanings to having fair skin. As to be shown in this chapter, Filipina YouTubers and their promotion of underarm whitening broker and perpetuate a dominant standard of beauty, femininity and flawlessness, which have been integral in the lexicon of Philippine media.

In the next section, we extend our discussion of gendered, racialised and classed representations of femininity through a neoliberal and postcolonial space of the Internet. We highlight the portraiture of an individualised, entrepreneurial as well as postcolonial femininity in Philippine society.

We also advance how the mechanisms of brokering are weaved into the ambivalent nature of neoliberal branding cultures and postcoloniality.

The Currency of a Lighter Skin through a Postfeminist Lens

Media texts and popular culture are powerful outlets for constructing ideal standards of femininity (Banet-Weiser, 2012; McRobbie, 2004). Girls and women are typically encouraged through traditional (Banet-Weiser, 2012; McRobbie, 2004) and new media (Kanai, 2019) to invest in improving their bodies (Strengers & Nicholls, 2017), enabling transformations and modifications that articulate empowerment (Gill & Scharff, 2011, Banet-Weiser, 2015). However, practices on achieving beauty, self-esteem and accessing a web of capitals is linked to consumerism. One has to purchase and use certain personal products to partake in routine care of the body. In a sense, the body becomes a site for constant monitoring and policing, which practices are shaped by gendered (Gill & Scharff, 2011; McRobbie, 2004) and commercial structures (Banet-Weiser, 2015). In this chapter, through the lens of a postfeminist critique of brand cultures, we inter-rogate how representations of bodily enhancements with skin whitening products among Filipina YouTubers contribute to the paradox surrounding the brokering of Filipina subjectivity. A salient point in our discussion is how representations occupy digital spaces and more likely demonstrate the negotiation of colour consciousness for visibility and commodification of gendered subjectivities. Notably, we focus on the case of the saleability of underarm whitening for women, a market that has been heavily targeted by underarm whitening brands (Baldo-Cubelo, 2015; Rondilla, 2009) as opposed to Filipino men who are typically targeted by companies that sell underarm deodorant or other skin care products (Lasco & Hardon, 2020).

Postfeminism highlights a set of ideologies, practices and strategies that often position women as individualised and entrepreneurial beings especially in neoliberal societies (Banet-Weiser, 2012; Gill & Scharff, 2011). In popular media, women are represented to engage with diverse consumption habits to achieve success, stability, mobility, and control in one's life (McRob-bie, 2004). This approach has been adopted across marketing campaigns of products for girls and women. Female consumers are presented with notions of empowerment through consumption of goods and services (Banet-Weiser, 2012). In contemporary times, digital spaces are explored and utilised by marketing companies to prescribe ideals of femininity (Banet-Weiser, 2012), which result to contradictory practices and affects (Kanai, 2019).

We are particularly interested in how ideal femininity is brokered in the commodified space of the Internet. We illuminate this by showcasing how YouTubers, just like brokers (Stovel & Shaw, 2012), collect, curate and disseminate relevant information to forge and sustain relations and capital building. However, we interpret the outcomes of these practices are not devoid from tensions. For this reason, we deploy the conceptual framing of "branding postfeminism" by Banet-Weiser (2012). For Banet-Weiser (2012), digital spaces have provided women to not visibilise or broadcast their everyday practices. However, digital practices have also been appropriated for brand cultures, compelling girls and women to perform based on hegemonic gendered discourses. For instance, in investigating how girls use YouTube to curate and present their everyday lives, Banet-Weiser (2011) argues that video production and a range of contents are largely influenced by adherence to gendered structures in order to generate hits, likes and shares. This means that authenticity is shaped by existing and powerful structures, such as the norms in digital cultures that stir high engagements among viewers (Banet-Weiser, 2011). Furthermore, online performativity is often prescribed to be aligned with branding an individualised and entrepreneurial self (Banet-Weiser, 2012; Duffy, 2017). As such, while these women utilise their own brand to create presence and intimacy in online spaces, they are also subject to cultural and commercial structures and conditions that often compel them to negotiate their capacities and performativity. It is for this reason that commodified and networked spaces are characterised for their ambivalent nature (Banet-Weiser, 2012; Glatt & Banet-Weiser, 2021).

In the context of our study, we pay attention to the use of online media for the brokering of certain ideals on femininity and beauty. Echoing Banet-Weiser's contention (2012), girls and women use digital media channels to create their brand in neoliberal and networked spaces. Often, enacting visibility, developing social networks and establishing intimate connections are achieved through branding strategies (Abidin, 2015; Duffy, 2017; Marwick, 2013). Branding strategies engage with gendered performativity, emphasising the construction of independence, an entrepreneurial vibe, and resilience despite the odds. Of a particular focus in this study though is the intertwining of gender, race and class in the performativity of constructing feminine subjectivity, complementing studies on how people of colour represent themselves in online spaces (Sobande, 2017; Nielsen, 2016). In our study, we critically reflect upon how postfeminist ideologies can be extended in the enactment of cultural whitening (Arnado, 2019) or practices through which Filipino women aspire for and enact a white aesthetic to navigate the uneven terrain of digital and commodified spaces. This point echoes studies that highlight how social hierarchies are well positioned in digital spaces, such as when brands promote whiteness through the use of a mobile application (Glenn, 2009) or the production of forums that cultivate ways of achieving Anglo features (Rondilla, 2009). Ultimately, women, especially from the Global South, or from the Philippines, negotiate brand and digital cultures, which we argue are influenced by colonial ideologies and histories. Further, "colour consciousness" manifests in online performances, interactivity from the audience, and even amplified by the technical features of the online platform (Jamerson, 2019) such as algorithmic recommendations (van Dijck, 2013).

In engaging with postfeminist critiques of brokering ideal femininity, we unearth the nuances and contradictions of representations generated through online practices and strategies. To achieve this is to draw the connections between the visualisation of beauty regimens and colonial history of the Philippines. We draw from the work of Bhabha (1994), noting how hybrid identities of colonial subjects engage with practices of "mimicry" to aspire for a dominant and privileged position occupied by colonisers. Here, colonised subjects desire and fantasise a sense of inclusion, status, and modernity by "mimicking" the practices of the colonisers. In the context of enacting white aesthetic, individuals aspire for and work their way in for achieving a lighter or white skin to embody whiteness. This act is sustained by internalising the positive meanings ascribed to whiteness, such as embodying purity, joyfulness, privilege, and high social status. However, mimicry has its excess, slippage, and ambivalence. The processes and practices of embodying an imagined whiteness tend to be defined as "white but not quite" (Bhabha, 1984). One may consistently attempt to change one's physique via the use of skin whitening or undergoing cosmetic surgery, but the outcome only serves as a resemblance.

Pinpointing the limits and ambivalence of mimicry is not only the concern of this chapter. We argue that mimicry is utilised as a branding strategy among Filipina YouTubers who monetise and benefit from their narratives and representations of ideal beauty. Certainly, Filipina YouTubers tend to position themselves as entrepreneurial and self-motivated in navigating digital and consumerist cultures. However, the ambivalence of representations manifests when transformations are "not quite" or may indicate what Arnado (2019) notes as temporary cultural whitening. The ambivalence also manifests especially when performativity is constructed by hegemonic ideals of beauty in the commodified space of the Internet. Nevertheless, this chapter showcases the brokering of ideal femininity among Filipina YouTubers who capitalise on their body care routines to achieve a whiter underarm. We now present our discussion of our data methods and analysis.

Data Collection

This chapter investigates ten videos produced by female YouTubers from the Philippines who present and curate tips and procedures on underarm whitening. In August 2020, we utilised Google trends to identify the most searched word on YouTube in the Philippines, particularly focusing on the topic of skin whitening. Instead of searching generic keywords "skin whitening," we opted for a more specific and local term. In such an approach, we used the Filipino word *pampaputi* or "whitening." After conducting the word search, we were provided with different phrases, such as *pampaputi ng kili-kili* (underarm whitening), *pampaputi ng balat* (skin whitening), and so forth. The keywords *pampaputi ng kili-kili* were the most searched keywords. From here, we used the keywords *pampaputi ng kili-kili* on YouTube search.

By keying in the words *pampaputi ng kili-kili* to search for videos on You-Tube, we accessed a total of 40 videos. These videos were collected based on accessing a main video as well as the recommended videos. The videos were filtered based on the number of subscribers and viewers. We only included videos with a minimum of 4,000 view hours and videos produced by YouTubers who have a minimum of 1,000 followers. This decision was made following YouTuber threshold on monetised content. Further, we selected videos produced by non-celebrities or by media institutions. We also excluded videos that do not focus on underarm whitening. Upon ranking the videos based on views, the top 10 videos were chosen (see Table 3.1). These videos were watched multiple times. Notes were produced to jot down key observations. The videos were transcribed and analysed. The four key frames used particularly looked at content, discursive style, credibility building, and platform specific strategies.

This chapter opted to employ anonymity for the data to protect the privacy of individuals in interracial relationships. We followed the ethical decision-making and recommendations of the Association of Internet Researchers (2012, 2019) by ensuring a careful and ethical handling, analysis and presentation of the selected and publicly accessible videos. We de-identified the names of the YouTubers by using pseudonyms. Following the work on Lange (2014) on protecting the privacy of content creators, we also used a generic title for the videos. Further, we also translated the quotes to English based on context and avoided incorporating texts as word-for-word (Zimmer, 2010). These approaches allowed us to protect the privacy of the YouTubers.

During the analysis for this chapter, we linked our observations to our key inquiry on unpacking how the ideal Filipina beauty is brokered. It is through this approach that we critically engaged with how the YouTubers negotiate postcolonial and marketised realities in a digital environment.

Table 3.1   Lifestyle brokers and their videos promoting underarm whitening

Broker	Date of joining Youtube	Subscribers	Number of videos	Title of video analyzed	Date published	Views
Karla	23 March 2016	1.62M	256	Underarm routine/ How did I whiten and smoothen my underarm.	12 October 2019	2,390,935 views
Carlene	16 February 2017	23.4K	57	How did my underarm whiten using baby oil.	27 March 2019	1,925,824 views
Andrea	16 April 2014	49.8K	676	How to whiten the underarm in a natural process.	5 January 2018	1,671,973 views
Susan	26 November 2018	35.3K	61	Cheap and Effective Underarm Whitening/ Underarm Routine	2 May 2020	1,376,540 views
Anna	22 October 2015	195K	172	My 3-day journey underarm whitening/ How to whiten the underarm in less than 3 days))	5 August 2019	1,056,016 views
Lisa	4 May 2016	39.6K	202	How did my underarm whiten	30 November 2019	974,450 views
Maria	8 May 2014	2.36M	89	Dark underams? Tips and tricks	4 October 2019	893,584 views
Molly	20 January 2017	36K	161	Effective and cheap underarm whitening	29 May 2019	890,838 views
Gale	16 June 2012	34.5K	212	Cheap underarm whitening/Super effective	2 November 2018	808,219 views
Athena	5 October 2015	109K	103	Cheap, thrifty and effective	22 April 2020	800,548 views

Prescribing Ways to Whiten the Underarm

The main topic of the ten YouTube videos was underarm whitening. Ten women shared their personal experiences and strategies in achieving a smooth and white underarm. They also highlighted the symbolic capitals that accessed as a reward of their persistent bodily care practices. We argue that their brokering practices contribute to the construction of the meanings and standards of ideal Filipino femininity, as often discursively represented in media contents (Baldo-Cubelo, 2015). In our investigation, these curated skin-related personal problems and successful transformation primarily brokered beauty ideals. In the following sections, we present the four dimensions of brokering that contribute to the articulation of beauty standards, including affective content, discursive style, credibility building, and platform-specific strategies.

The Travails of Skin Transformation

Most of the videos structured the process of underarm whitening through a logical sequence of skin transformation. We argue that this approach tends to complement the approach of aggressive marketing campaigns and advertisements on skin improvement – showing the problem, finding and applying the solution, and activating a renewed and glowing skin (Baldo-Cubelo, 2015; Rondilla, 2009). This step-by-step process essentially allows the YouTubers to generate value to and legitimise their beauty regimens.

Firstly, the YouTubers spoke about their issue – a dark underarm. They highlighted a dark, rough, hairy, and sometimes smelly underarm. Such conditions were abhorred, highlighting the existence of bodily issues that counter the aesthetic of beauty ideals such as having bright and flawless underarms. Furthermore, they mentioned the impact of having a dark underarm on their confidence to wear a range of clothes. A case in point is reflected in the statement of Anna, "It's stressful because I used to have dark underarms and I know that some of you are going through with the same feeling that you can't wear your favourite sleeveless because you're not overconfident with your underarms."

Secondly, YouTubers highlighted the cause of hyperpigmentation. These women blamed shaving, plucking, and aggressive scrubbing for the darkening of their skin. In some cases, four YouTubers pinpointed hormonal changes during pregnancy as a culprit of their skin issue. As Anna shared:

I started having dark underarms when I got pregnant. It's not only the underarm but also my nape, groin area, and other body parts, and even the hidden parts of the body, which is normal because of the hormonal changes in the body during pregnancy. So that's the problem. Everything returned to its condition except for my underarms and my underarms became darker…

Identifying the causes of a dark underarm also led the YouTubers to prescribe caring practices for one's body, including avoiding shaving, aggressive scrubbing and even plucking. For instance, Maria provided a list of solutions to avoid the darkening of the underarm. She suggested to her viewers to stop shaving. She also highlighted the use of a deodorant with whitening ingredients. Moreover, she advised her viewers to avoid wearing clothes with ruffles which can irritate the underarm. But more importantly, she highlighted the importance of caring for one's underarm during day and night. For her, bodily care must be worked on from morning until night. As such, she mentioned a product and explained its effect.

For me, it's very important that if you use something (referring to under-arm whitening) in the morning, then you should also apply something at night because your body is relaxed when you sleep. So for example in skin care, when we take care of our skin, we put a lot of things and it's even better if you care so much at night because your skin takes a rest and it will absorb the product.

Essentially, the YouTubers suggested that women like them must reclaim their bodies that were undermined by their own aggressive skin practices. Here, bodily issues become a self-responsibility that demands urgent action (Elias, Gill, & Schraff, 2017).

One of the major concerns in altering the underarm into a white and smooth state is the lack of access to affordable yet effective products. In the Philippines, studies have shown that expensive skin products can deter Filipino consumers from buying safe products and opting for informal markets (Mendoza, 2014; Ng & Lachica, 2020). However, over the past years, YouTubers have become an active and major player in promoting alternative and relatively affordable skin whitening products in the Internet, with some even showing injectables for skin lightening (Ng & Lachica, 2020). In our study, the YouTubers served as brokers in finding ways to address the issue of accessing affordable underarm products. They provided and recommended a range of alternative, accessible, affordable, and claimed to be effective products. They even demonstrated how the product is used and visibilised its outcome.

In a way, this approach countered issues of costs especially for women who could not find a cheap yet effective underarm whitening product. Firstly, some YouTubers featured the use of readily available and cheap skin care products in the market. This was the case for most YouTubers who typically reviewed products and other goods. In some cases though, existing personal care products and natural remedies were recommended. For instance, Carlene admitted that she's a "budgetarian" or a budget conscious person. However, despite limited financial resources, she aspired and worked on her way to achieve white underarms. She used body oil. As she confessed, "I am a budgetarian mother. I tried it [*body oil*] and yes it was very effective."

Secondly, others showed the production of a do-it-yourself mixture which was transferred to a spray bottle. This was the case of Andrea, who demonstrated how to create a do-it-yourself underarm whitening product. In the video, the YouTuber prepared and mixed three key ingredients – water, deodorant crystals (*tawas*) and lime (*kalamansi*) (see Figure 3.1). The mixed ingredients were put in a spray bottle, which the YouTuber used on her underarm. Andrea was very proud of her affordable alternative for an underarm whitening product, "You already have a deo-spray for 5 pesos for the lime, 15 pesos for the deodorant crystals, water, and spray bottle."

Third, a YouTuber like Karla opted for an alternative, accessible and cheap option for underarm laser treatment. In her video, she narrated that she inquired in a clinic about the cost of a laser treatment for the underarm. She then learnt that the cost was 5,000 pesos (approx US$95). As she did not have the budget, she decided to get a similar and cheaper product online. Now, she uses this product to lighten her underarm. As she said, "I just had a few sessions and it's quite economical. It really looks like the tools used in a clinic. It's money saving if you'll use it at home. This tool can be used in the house and this device can be used for the legs and armpits, upper lip, hairline, bikini line, and anywhere."

Overall, YouTubers demonstrated that there are alternative options to address one's dark underarm, contributing to the brokering of beauty standards. Indeed, this performativity shows that women are left with no excuse in improving their body because the YouTubers showed that one can be practical and strategic in achieving a glowing and smooth underarm. Notably, on top of the concerted and practical efforts to embody a white underarm, one is encouraged to be "patient." As Karla said, "My advice to you is you need to be patient especially at a time when there is no instant.

Figure 3.1   Still from a YouTube video. The ingredients to be used for a DIY
              underarm whitening product.

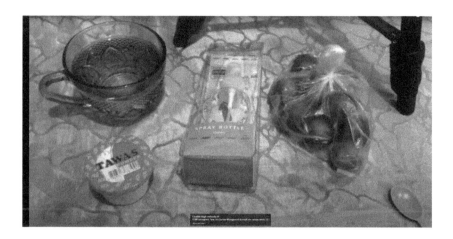

Whatever will that be, work hard for it. Whatever is that, like a job, person,
or an underarm. Your problem could also be a similar problem to others.
You just need to find products that suit you."

Tropes, Tools, and Tactics

Most of the videos were delivered through a genre of confessional talk
(Raun, 2018). The YouTubers mostly faced and spoke to the camera. The
"storytelling" was also sometimes supported by visuals of a dark underarm.
As the videos focussed on featuring the underarm, most YouTubers wore
sleeveless. Evidently, the outfit signalled the revelation of a whiter and
rejuvenated underarm at some point of the narrative. As the storytelling
continued, the YouTubers revealed a white underarm. This underarm also
served as the surface when the YouTubers demonstrated the application of
the whitening cream or liquid. Ultimately, the confessional talk was paired
with the spectacle of a "before" and "after" visual of the "problem" area. This
structure guided the viewers in bodily transformations.

   In marketing campaigns of skin whitening products, having a dark skin
tone is considered a problem (Baldo-Cubelo, 2015). The consumers are made
to feel guilty and then encouraged to act on this problem by buying and
consuming the right and effective product to address the skin problem and
eventually boost one's confidence and status in life (Baldo-Cubelo, 2015;

Mendoza, 2014; Lasco & Hardon, 2020). In the digital space, YouTubers used a tactic to highlight the underarm issue and enticed the viewers to watch and follow their transformative journey. Notably, for the YouTubers, different discursive "monikers" were used to brand a dark underarm. For instance, Anna referred to her dark underarm as *madilim na nakaraan* (dark past). Gale called it *makulimlim* (shady). Andrea used the term *mashoho* (stinky). These words reflected the internalised abhorrence for a dark underarm that demands immediate action through the consumption of various products.

YouTubers brokered the "tools of the trade" in bodily enhancements. Whitening creams, skin care products, and natural remedies were showcased in the videos. We noticed that skin care products like body oil, cotton balls, and lime were prominent in several videos. In several do-it-yourself demonstrations, the YouTubers showed how each ingredient was prepared and combined to produce a mixture (see Figure 3.2). Graphics were juxtaposed with the shots of the ingredients. In cases that existing products are used, the YouTubers link the product information in the description box. For instance, the description box of Ariza had the following texts: "PRODUCTS MENTIONED: Johnson's baby oil PHP 25, Luxe organic aloe vera gel PHP 199 [link of Luxe], Milcu deodorant powder PHP."

Thumbnails were utilised to show the highlight of the video. Majority of the videos had a photo of the YouTuber with a light or white underarm, a photo of a dark underarm, some products, and a catchy phrase. In some videos, a thumbnail highlighted a "fast" solution for addressing dark underarms. The videos of Susan had *Paano pumuti in just one week* (How to whiten in just one week). Anna also claimed, *Maputing kilikili in less than 3 days* (White underarm in less than 3 days). These time-based statements were presented to attract the viewers in accessing the different and quick steps in achieving a whitened underarm. However, to prove the effectiveness of using products for a quick transformation, YouTubers still showed how the products were accessed, concocted, and used. For instance, in the case of Anna, the different ingredients in creating a DIY product for whitening the underarm were showcased. She was seen mixing and applying the mixture on her underarms. Interestingly, she mentioned that her demonstration was based on a YouTuber who showed how to whiten the underarms in three days. She thanked the YouTuber in the description box of the video: "Thanks to *LovelySkin* for the inspiration!" In the video, she used graphics to highlight which video clip referred to day one and day three. In day one, she put the mixture on her underarm. On day three, she revealed her underarm. After showing a before and after shot, she said, "So there you go and I am feeling shy in showing my underarms but at least I can be proud

Figure 3.2   Still from a YouTube video. The process of preparing a DIY underarm
              whitening product.

because there was an improvement and I can see a clearer and improved underarm from my dark underarm. I am happy because it's summer here so I can wear the sleeveless that I used to wear before."

The tips for whitening the underarms were framed as a form of "help" by the YouTubers to their viewers who might be struggling with underarm issues. The statements often highlighted that the YouTubers went through experiences and these might be similar to viewers. As such, they were positioned as women who are helping other women with their bodily issues. In here, a sense of collective emerged but particularly for the embodiment of dominant beauty ideals. For instance, Carlene highlighted, "I hope this video helped my 'sis' out there who are budgetarian or thrifty like me." And on a more general term, a YouTuber like Anna emphasised, "Hopefully it (*referring to the video*) helped those who are going through a gloomy future with their underarms." In such a statement, the value of the video is framed as a source of collective knowledge and skills among women who want to address a "dark" future because of having skin issues.

Lastly, YouTubers capitalised on interactivity to engage the audiences. They purposely give in on the viewers' demands to craft their content, reflecting relational labour (Baym, 2015). For instance, Maria mentioned in her video that her viewers saw her photo while she was in a popular beach destination in the Philippines. In the photo, her white underarm was seen. She then received some comments, asking about her routines in having a white underarm: "I had a photo showing my arms up on the beach and then many of you commented, Ate Maria, can you share your tips on your underarm. What do you do with your underarm? Why is your underarm white?" This statement prompted her to produce a video, "So this video is the answer to all of your answers." Meanwhile, some are motivated to produce videos through direct video requests. This was the case for Athena, saying, "This is the most requested video in my channel. Many commented in the comment section that I produce this video." Nevertheless, these both statements show that the audience or viewers are curious about some tips and procedures in whitening the underarm. As such, the performativity of the YouTubers adjusts to the viewers' requests in order to sustain engagement and more likely generate profit.

Building the credibility of a skin guru

The ten videos capitalised on the aesthetic of transformation to establish the credibility of using an existing or a DIY underarm whitening. YouTubers typically showed a close-up photo of their dark underarm and this is typically contrasted to the final result. A skin revelation was followed by a personalised, playful and intimate discussion of skin regimens. The narratives, which centre on individual experiences, act as mechanisms for brokering beauty ideals and the perceived benefits.

There are different ways through which the YouTubers establish the effectiveness of their product. Firstly, several of them used a timeline approach. This means that the video presented a compilation of underarm whitening routines through visuals and storytelling, leading up to a final and whitened state of the underarm. A case in point is Molly who demonstrated the creation of her affordable DIY version of *tawasmansi* (deodorant crystals with lime), an underarm whitening mixture. After producing the mixture, she started applying it on her underarm. The videos presented the transformation of her underarm during the third and tenth day of use. So far, she was happy with the results. However, Molly also articulated a sense of ambivalence in the final outcome of the transformation. She still identified some dark shades in her underarm. As she shared:

> I am so happy with the results! Right now this is the status of my underarm. The chicken skin slightly vanished and lightened. The dark shade, this particular line, it's quite difficult to remove that fold in the underarm. That's the number one problem in our underarm, the fold. But all in all I am super happy, in the past, in my first update, I had dark shades in my chicken skin then, right now it's more lightened.

For some who did not use a timeline approach, the current state of a whitened underarm is shown to the camera. The YouTuber then talked about the products and routines that helped the further whitening of the skin. This was the approach by several YouTubers like Lisa, who tried using underarm whitening products to restore the condition of her underarm prior to pregnancy. In the video, Lisa admitted that she had gone through 15 sessions of laser treatment to make her underarm smooth. On top of this, she detailed in the video her routine to maintain her white underarm. She presented a step-by-step demonstration of applying different products, such as body oil, a spa milk salt, a lime, and also an Aloe Vera gel. As she explained the use of these products, she also warned the viewers who want to try the routine to pay attention to their skin's sensitivity to avoid irritation.

Most of the YouTubers warned their viewers to be extra careful in using a range of products to whiten and smoothen the underarm. As brokers, they highlighted the downside of using a range of skin whitening products. The word "irritate" was often used to emphasise a negative outcome. Yet, despite raising their concerns over product use, they still promote and use a range of underarm care products. To protect themselves from potential complaints from their viewers who might follow their regimens, they used a "disclaimer" in recommending a range of products, services and practices. Among the ten YouTubers, nine used a disclaimer. Below are some examples of the YouTubers' disclaimer:

> The products that I'm using MAY OR MAY NOT WORK for you. We have different SKIN TYPES ✨✨ ☺☺

> Any information associated with this video should not be considered as a substitute for prescription of any professionals. Please always remember that we are not all the same. Products that work on me, may or may not work on you. "USE IT AT YOUR OWN RISK"

> My kilikili is not perfect (*My underarm is not perfect.*). I am not a skin expert/dermatologist.

A "disclaimer" operates in two ways. Firstly, it allows the YouTuber to detach oneself from the viewer who might use the recommended product and goods. This means that the YouTuber is not accountable for the potential effects of the featured underarm whitening cream or liquid spray. Secondly, the disclaimer highlights the specific socio-cultural reading of the suitability of a certain product to individuals. According to Hardon (1992), Filipinos tend to assess the efficacy of medicines in everyday conversations through assessments of suitability or *hiyang*. *Hiyang* is understood by comparing the drug's effects on oneself or others as well as effects based on dosage level and product price (Hardon, 1992). For the former, if the product does not work for oneself but works for others, then the product is not *hiyang* for oneself. For the latter, if a high level of dosage still does not work for oneself, then one should stop taking it because it is not *hiyang* for oneself. As such, *hiyang* becomes the basis for Filipinos to opt for alternative cure or therapy. When a drug does not meet the expectations of the individual, instead of "letting" the medicine work, other options are considered. Notably, it is also important to point out that the lack of access to health services and medication compel Filipinos to opt for alternative solutions such as herbal medicines or buying drugs without prescriptions.

In our analysis, the narratives of the YouTubers become prompts for the viewers to assess the efficacy of the whitening products. Viewers were informed that the products or services accessed by the YouTubers are *hiyang* for them. It is through this point that the "disclaimer" works. The YouTuber can detach oneself from any risks or tensions that the viewers may experience in using their promoted products or services because they reiterated in the disclaimer that the product is *hiyang* for them and may or may not be *hiyang* for the viewers. Despite this uneven effect, the videos construct the imagined efficacy of the skin-whitening product via displaying a white and smooth underarm. This outcome then becomes the basis for legitimacy of the product, therefore attracting the viewers to watch the video, follow the YouTuber's routines, and even use the same products to enact transformation.

Lastly, YouTubers who claimed a fast and effective way of underarm whitening presented some documentation of the transformation of the underarm. This was the case for Susan, who showed how to whiten one's underarm in "just one week." Interestingly, Susan did not show a sequence of before and after images. Rather, she presented her routine, such as using body oil and lemon to cleanse and whiten her underarm. She demonstrated how to apply the liquids on her skin. Further, she also promoted ways of taking care of one's underarm, including avoiding plucking and shaving, opting for waxing, the use of *hiyang* (suitable) antiperspirant or deodorant, and gently scrubbing the underarm. These practices were highlighted to ensure the whitening and maintenance of the underarm.

The textures of platform-specific strategies

The ten videos contained different platform strategies to ensure a sustained engagement of the imagined audiences. Hashtags were common in four out of ten videos. Hashtags were embedded in the upper section of the video title. Examples were *#Kilikiliroutine* (Underarm routine), *#MurangPampa-putiNgKilikili* (Affordable underarm whitening product), #lightenarmpits, #darkspots, #natural remedies, and so forth. In some cases, a YouTuber also took pride in representing oneself through a hashtag. For example, Molly used the hashtags, *#FilipinaYouTuber*, *#DIYTawasMansi* and *#Kilikiligoals* (underarm goals). Using the hashtag *#DIYTawasMansi* primarily highlights the "TawasMansi" or Potassium Alum mixed with lemon.

The description box was also filled with relevant information. Texts often include the products used in the demonstration, the links where the used products can be accessed or bought, the ingredients used in producing a DIY underarm whitening liquid, the YouTuber's social media handle, an email for interested businesses and partners to advertise and collaborate, as well as other online channels for selling products and services. For example, in the description box of Molly's video, a range of information was shown, including a disclaimer, a list of links of past YouTube videos, promotion of another YouTuber, an email for collaboration, hashtags, and tools used for the production of the video.

> Hello! 🙈😳 DIY TAWAS MANSI. Hope it helps you 😊😊 (Don't Judge 😋) Please maximize or minimize the volume if needed. Sorry about my audio. Thankyou! 😳😳 DISCLAIMER: The products that im using MAY OR MAY NOT WORK for you. We have different SKIN TYPES ✨✨ 😊😊
> Product Mentioned: Water Empty Spray Bottle Tawas Calamasi
> You can also watch: How i get Monetised: [YouTube link]
> My First YouTube Sweldo: [YouTube link]
> Filming Set Up for Beginners: [YouTube link]
> My Glow Up Story: [YouTube link]
> My Weight Loss Journey: [YouTube link]
> KILI Story Part 1: [YouTube link]
> Achieve Glass Skin (Skincare Routine): [YouTube link]
> PLEASE DO SUBSCRIBE ON MY NIECE YT CHANNEL: [YouTube link]
> Follow me on: Tiktok @Molly
> Facebook [Facebook profile page]
> Instagram [Instagram page]
> FOR BUSINESS/COLLABS/OTHERS Email me at: (email address)

Camera Used: Samsung NX Mini/Iphone 6+
Video Editor: iMovie
Thumbnail Editor: Picsart
Thank you for watching!!!!! 💤💤 ILOVEYOU GUYSSS 😍😘😙😚 #Filipi-
naYouTuber #DIYTawasMansi #KILIKILIGOALS

The last few frames of each video also reflected platform-specific strategies. In some instances, YouTubers showed their social media channels and invited their viewers to connect and follow them in these outlets. But commonly, the last frame of the video was embedded with thumbnails of the YouTuber's past videos. In here, the algorithms worked by feeding viewers with similar types of contents on beauty routines. For instance, in the video of Andrea, suggested videos related to her underarm whitening video were about her previous video on trying a new make-up set and buying the skin whitening product glutathione. Similarly, a series of recommended videos appeared in the last frame of Gale's video, presenting titles on beauty routines, including permanent underarm hair removal, whitening of the knees, elbow, and other parts of the body, and an explanation on sperm facial mask. Nevertheless, these recommendations were operationalised based on coding beauty regimes of the YouTuber contributing to the formation of filtered aesthetic and ideals of femininity.

Lastly, YouTubers deployed call-to-action statements to ensure the sustained consumption of their viewers of their online contents. For example, Carlene said in her video, "Please don't forget to like this video. If you do, don't forget to subscribe and click the bell button so you can be updated with my uploads. Thank you guys. Bye." After delivering her line, the video transitioned to some graphics stating, "Thanks for watching. God bless." These texts were accompanied by a kiss emoji. In some cases, some YouTubers encouraged a more participatory engagement. For instance, Andrea, who showed a step-by-step procedure of a skin whitening liquid in a spray bottle, was encouraging her viewers to comment on her video especially among those who would try her prescribed skin whitening solution. As she said, "So that's it for our video, the first DIY of the year. Thank you for watching. Don't forget to like and subscribe to my channel. And please comment if you will also try it. Bye bye." Andrea's approach in engaging audiences was similar to Karla's strategy. However, Karla was performing more of a relational labour (Baym, 2015) by positioning herself as readily available to respond to any questions from her viewers. As she noted, "If you already tried these products, just comment down below and share your experiences with me. And if you have any questions, just leave a question or comment in the comment section and I will try to answer them." Here, one can interpret this as a strategy to know the impact of the product to the user as well as boost potential engagement from the audience through comments.

For some YouTubers, the rhetoric of care was used to encourage viewers in consuming and sharing the videos. This gesture may contribute to a sustained circulation and consumption of the YouTuber's video, which then drives the monetisation of content as well as reinforcement of the aesthetics of beauty ideals. This was the case of Susan, stating:

> Thank you for watching. I hope you learned something on this day. Sharing is caring. Why don't you share this video to your friends so they can also brighten their dark underarm. If you do like this video, don't forget to give me a thumbs up. And thank you so much again for watching. Don't forget to subscribe. Click the bell button so you can be notified if I have a new uploaded video. So you won't miss out any tips that I will be sharing to everyone. Thank you very much.

In Susan's statement, the words "sharing is caring" is highlighted. This means that although whitening one's underarm can be considered an individual journey, the potentiality of effecting positive change for others through collective and networked practices can be enacted upon. One becomes a pillar of support for others who have an underarm problem, which solution can be addressed through "recommending" the YouTuber's content on underarm whitening. In a sense, once an individual views the video, the products are promoted, contents are continuously recommended, and beauty standards are reinforced through a loop of contents operationalised and controlled by algorithms.

Commodifying a Loop of Beauty Ideals

Based on the ten videos that we analysed, we observed how contents brokered the aesthetic of beauty ideals for Filipino women. The YouTubers build on their personal experiences to collate and distribute useful information to enact an affordable and effective skin transformation. In the first instance, the personalised narratives in a confessional format talked about the problems of and solution to a dark underarm. In the brokering process, they highlighted their frustrations, skin care practices, and successful transformation. These diverse themes in content creation were weaved together to capture, curate and perpetuate dominant ideologies on being an empowered, strong and confident Filipino woman, which has been a common framing in traditional media channels (Baldo-Cubelo, 2015).

Despite the emphasis on ways of achieving a white underarm, the YouTubers deployed a disclaimer. This means that what works for them might not work for others. In the context of Philippine culture, the disclaimer from the YouTubers can be acceptable especially when medicinal products are assessed based on *hiyang*. As such, the YouTubers are emphasising that there are different effects of the products on individuals, but one can still attempt to try for such products can be *hiyang* for oneself. And there's only one way to determine the suitability of the product, which according to the YouTuber, is to verbalise and share one's experience in using similar products through commenting on the video.

As brokers, these YouTubers promote beauty ideals, specifically in embodying their smooth and glowing underarms. The visuals created the desire for whiteness, highlighting that women who have dark underarms by birth, as a result of pregnancy, or "improper" skin care practices can restore or have a white underarm. Here, the dark underarm was abhorred and only the use of commercial and DIY skin care products can solve the problem that is essentially skin deep – lack of confidence. Through a postfeminist lens, the YouTubers operate as conduits in the amplification of postfeminist branding (Banet-Weiser, 2012). For the female YouTubers, bodily issues were used to engage audiences and paved the way for entrepreneurial practices. Further, they embodied a sense of colour consciousness, highlighting the capitals that go with having a white underarm – confidence, beauty, and admiration from the viewers. The viewers were invited to watch the videos in the hopes of enacting skin transformations on themselves and accessing various capitals – a sense of empowerment, beauty, and privilege.

A sense of collective was harnessed by YouTubers in digital and feminised spaces. Sharing of tips and tactics for beautification means caring for others who might be going through similar problems. This, we argue, is another co-optation of postfeminist branding. In digital cultures, the collective matters because it fuels hits, views and subscriptions. For online content creators, the collective is paradoxically harnessed for entrepreneurial quests. These individuals highlight their relatable problems, generating intimate connections from viewers (Abidin, 2015). In some cases, audiences demanded from the YouTubers to cover a pressing issue – whitening the underarm or any other lifestyle-related topics. In here, the audiences suggested topics that internalise gendered, racialised and classed hierarchy, which is then appropriated by the YouTuber as a marketable content. As such, it seems that the collective is embedded in the ambivalence of brand cultures – enabling empowerment through visibility and participation yet cementing dominant discourses. Significantly, while the videos showcased how to improve one's body, hegemonic discourses are constantly negotiated through platform-specific modality. In digital spaces, the production and consumption of hegemonic representations are more likely perpetuated by a loop of contents that are influenced by algorithms which often circulate contents based on similarity and popularity (Pariser, 2011; van Dijck, 2013). In this case, YouTube, as we conceptualised in Chapter 2, serves as a socio-technical broker that circulates information based on the conditions enacted by the content creators, such as the production and distribution of creative contents that are mostly viewed, liked, and shared by viewers, and such practices fuel a loop of similar or matching and discursive contents. Nonetheless, YouTube becomes a space where hierarchical discourses in the Philippines are perpetually negotiated through a set of commercial practices, networked arrangements, and internalised colour consciousness.

Beyond Skin Deep? On the Paradox of Digital Brokering

This chapter has presented the brokerage of beauty ideals among selected videos of Filipina YouTubers. It particularly highlights the range of practices and strategies that promote and enact the whitening and smoothening of the underarm. By closely examining the aesthetic, rhetoric and technical features of online contents, we have demonstrated how desires and practices of having a flawless underarm pave the way for negotiating hegemonic structures in Philippine society. One the one hand, the Filipina YouTuber utilises creative, playful and personalised contents to visibilise oneself and even harness a sense of collective. However, online performances tend to resonate existing marketing campaigns on feminine beauty and lifestyle (Baldo-Cubelo, 2015, Rondilla, 2009). The body, particularly having a dark underarm, signifies embarrassment and repulsion, and this affective state demands a solution such as consuming or creating underarm whitening products. As such, the online narratives particularly position empowerment through entrepreneurialism (Banet-Weiser, 2012). Moreover, we also argue that the videos essentially co-opt hegemonic racialised desires of having a bright and smooth underarm, which is positioned as crucial for accessing diverse symbolic capitals – confidence and status. Meanwhile, the viewers are then presented with texts and resources that tap into imaginaries and criteria of flourishing feminine subjectivity. They are also invited to partake in a sense of collective as imagined by the YouTuber, including acting on one's skin problem, and inviting others to address similar issues through networked practices. Notably, the technical organisation of YouTube also complements the loop of cultural whitening. Through algorithmic systems, the recommended videos align with the key themes of the videos produced by the YouTuber and potentially consumed by the viewers. Ultimately, we argue that the YouTube videos on constructing Filipina beauty standards, as amplified by platform specific strategies and technical organisation, indicate the negotiations of subjectivity, positionality, and visibility in postcolonial and neoliberal Philippine society.

References

Abidin, C. (2015). Communicative ♥ intimacies: Influencers and perceived interconnectedness. *Ada: A Journal of Gender, New Media, and Technology, 8,* 1–16. doi:10.7264/N3MW2FFG

Abidin, C. (2018). *Internet celebrity: Understanding fame online* (First ed.). Emerald Publishing Limited.

Arnado, J. M. (2019). Cultural whitening, mobility and differentiation: Lived experiences of Filipina wives to white men. *Journal of Ethnic and Migration Studies,* 1–17. doi:10.1080/1369 183X.2019.1696668

Association of Internet Researchers. (2012). *Ethical decision-making and Internet research: Recommendations from the AoIR ethics working committee* (version 2.0). Retrieved 12 January 2020, https://aoir.org/reports/ethics2.pdf

Association of Internet Researchers. (2019). *Internet research: Ethical guidelines 3.0.* Retrieved 10 January 2020, https://aoir.org/reports/ethics3.pdf

Baldo-Cubelo, J. T. (2015). The embodiment of the new woman: Advertisements' mobilization of women's bodies through co-optation of feminist ideologies. *Plaridel: A Philippine Journal of Communication, Media, and Society, 12*(1), 42–65.

Banet-Weiser, S. (2011). Branding the post-feminist self: Girls' video production on YouTube. In M. C. Kearney (Ed.), *Mediated girlhoods: New explorations of girls' media culture* (pp. 277–294). Routledge.

Banet-Weiser, S. (2012). *Authentic TM: The politics and ambivalence in a brand culture.* York University Press.

Banet-Weiser, S. (2015). Keynote Address: Media, markets, gender. Economies of visibility in a neoliberal moment. *Communication Review, 18*(1), 53–70. doi:10.1080/10714421.2015.996398

Baym, N. (2015). Connect with your audience! The relational labour of connection. *Communication Review, 18*(1), 14–22. doi:10.1080/10714421.2015.996401

Bhabha, H. (1984). Of mimicry and man: The ambivalence of colonial discourse. *Discipleship: A Special Issue on Psychoanalysis, 28,* 125–133.

Bhabha, H. (1994). *The location of culture.* Routledge.

David, E. J. R. (2013). *Brown skin, white minds: Filipino-American postcolonial psychology.* Information Age Pub. Inc.

Dixon, A. R., & Telles, E. E. (2017). Skin color and colorism: Global research, concepts, and measurement. *Annual Review of Sociology, 43,* 405–424. doi:10.1146/annurev-soc-060116-053315

Duffy, B. E. (2017). *(Not) getting paid to do what you love: Gender, social media, and aspirational work*. Yale University Press.

Elias, A., Gill, R., & Schraff, C. (2017). Aesthetic labour: Beauty politics in neoliberalism. In A. Elias, R. Gill, & C. Schraff (Eds.), *Aesthetic labour: Beauty politics in neoliberalism* (pp. 3–50). Palgrave Macmillan.

Gill, R., & Scharff, C. (2011). *New femininities postfeminism, neoliberalism, and subjectivity*. Palgrave Macmillan.

Glatt, Z., & Banet-Weiser, S. (2021). Productive ambivalence, economies of visibility and the political potential of feminist YouTubers. In S. Cunningham & D. Craig (eds.), *Creator culture: Studying the social media entertainment industry* (pp. 39–56). New York University Press.

Glenn, E. N. (2008). Yearning for lightness: Transnational circuits in the marketing and consumption of skin lighteners. *Gender & Society, 22*(3), 281–302. doi:10.1177/0891243208316089.

Glenn, E. N. (2009). Consuming lightness: Segmented markets and global capital in the skin-whitening trade. In G. E. N. (Ed.), *Shades of difference: Why skin color matters* (pp. 166–187). Stanford University Press.

Global Industry Analysts. (2020). Skin lighteners. Market analysis: Trends and forecasts. Retrieved 21 December 2020, https://www.strategyr.com/MCP-6140.asp

Hall, R. (1995). The bleaching syndrome: African Americans' response to cultural domination vis-a-vis skin color. *Journal of Black Studies, 26*(2), 172–184.

Hardon, A. P. (1992). That drug is hiyang for me: Lay perceptions of the efficacy of drugs in Manila, Philippines. *Central Issues in Anthropology, 10*(1), 86–93. doi:10.1525/cia.1992.10.1.86

Harris, A. (2009). Introduction: Economies of color. In G. E. N. (Ed.), *Shades of difference: Why skin color matters* (pp. 1–5). Stanford University Press.

Hunter, M. (2007). The persistent problem of colorism: Skin tone, status, and inequality. *Sociology Compass, 1*(1), 237–254.

Hunter, M. L. (2005). *Race, gender, and the politics of skin tone*. Routledge.

Hunter, M. L. (2011). Buying racial capital: Skin-bleaching and cosmetic surgery in a globalized world. *Journal of Pan African Studies, 4*(4), 142.

Jamerson, T. W. (2019). Race, markets, and digital technologies: Historical and conceptual frameworks. In G. D. Johnson, K. D. Thomas, A. K. Harrison, & S. A. Grier (Eds.), *Race in the marketplace* (pp. 39–54). Palgrave Macmillan.

Jha, M. R. (2016). *The global beauty industry: Colorism, racism, and the national body*. Routledge.

Kanai, A. (2019). *Gender and relatability in digital culture: Managing affect, intimacy and value*. Palgrave Macmillan.

Lange, P. (2014). Commenting on YouTube rants: Perceptions of inappropriateness or civic engagement? *Journal of Pragmatics: An Interdisciplinary Journal of Language Studies, 73*, 53–65. doi:10.1016/j.pragma.2014.07.004

Lasco, G., & Hardon, A. P. (2020). Keeping up with the times: Skin-lightening practices among young men in the Philippines. *Culture, Health and Sexuality, 22*(7), 838–853. doi:10.1080/13691058.2019.1671495

Marwick, A. E. (2013). *Status update: Celebrity, publicity, and branding in the social media age*. Yale University Press.

McRobbie, A. (2004). Post-feminism and popular culture. *Feminist Media Studies, 4*(3), 255–264. doi:10.1080/1468077042000309937

Mendoza, R. L. (2014). The skin whitening industry in the Philippines. *Journal of Public Health Policy, 35*(2), 219–238.

Ng, D., & Lachica, L. (2020). The dangers of trying to be fairest of them all in the Philippine. Retrieved 12 November 2020, https://www.channelnewsasia.com/news/cnainsider/dangers-trying-be-fairest-of-them-all-philippines-skin-whitening-12642522

Nielsen, M. (2016). Love Inc.: Toward structural intersectional analysis of online dating sites and applications. In S. U. Noble & B. M. Tynes (Eds.), *The intersectional Internet: Race, sex, class, and culture online* (pp. 161–178). Peter Lang.

Pariser, E. (2011). *The filter bubble: What the Internet is hiding from you*. Penguin Press.

Rafael, V. (2000). *White love and other events in Filipino history*. Duke University Press.

Raun, T. (2018). Capitalizing intimacy: New subcultural forms of micro-celebrity strategies and affective labour on YouTube. *Convergence, 24*(1), 99–113.

Rondilla, J. (2009). Filipinos and the color complex, ideal Asian beauty. In G. E. N. (Ed.), *Shades of difference: Why skin color matters* (pp. 63–80). Stanford University Press.

Rondilla, J. L. (2012). *Colonial faces: Beauty and skin color hierarchy in the Philippines and the U.S.* (Doctoral dissertation), University of California, Berkeley., Berkeley, CA.

Rondilla, J. L., & Spickard, P. (2007). *Is lighter better?: Skin-tone discrimination among Asian Americans*. Rowman & Littlefield Publishers

Sobande, F. (2017). Watching me watching you: Black women in Britain on YouTube. *European Journal of Cultural Studies, 20*(6), 655–671. doi:10.1177/1367549417733001

Strengers, Y., & Nicholls, L. (2017). Aesthetic pleasures and gendered tech-work in the 21st-century smart home. *Media International Australia*. doi:10.1177/1329878X17737661.

Stovel, K., & Shaw, L. (2012). Brokerage. *Annual Review of Sociology, 38*(1), 139–158. doi:10.1146/annurev-soc-081309-150054

van Dijck, J. (2013). *The culture of connectivity: A critical history of social media*. Oxford University Press.

Yip, J., Ainsworth, S., & Hugh, M. T. (2019). Beyond whiteness: Perspectives on the rise of the Pan-Asian beauty ideal. In G. D. Johnson, K. D. Thomas, A. K. Harrison, & S. A. Grier (Eds.), *Race in the marketplace* (pp. 73–85). Palgrave Macmillan.

Zimmer, M. (2010). 'But the data is already public': On the ethics of research in Facebook. *Ethics and Information Technology*, 12, 313–325.

4   Relationships

Abstract

This chapter illuminates the curation of interracial relationships between Filipino women and their foreign male partners on YouTube. Drawing upon the conception of countererotics, cultural whitening, and a postfeminist critique, it problematises intimate, affective and creative approaches enacted by YouTubers in networked and monetised environments as a form of brokering of idealised interracial coupling. The analysis of LDR (long-distance relationship)-themed YouTube videos is situated within the socio-historical context and technological landscape of mixed-race partnership and marriages in the Philippines. Ultimately, the emotive narratives, personalisation mechanisms, and platform-specific strategies construct and marketise the vernaculars of interracial relationships. The chapter ends with the possibilities, economies, and parameters of broadcasting intimate relationships in networked spaces.

Keywords: interracial relationship, mediated intimacy, postfeminist entrepreneurialism, Long-distance relationship, dating apps, marriage

Unprecedented industrialisation, global expansion of markets, and the rapid development of transport and communication technologies have facilitated new ways for the conduct of individual and family life. Sociologist Anthony Giddens (1991) has identified this arrangement because of individualisation in modern society, highlighting the changing notions and practices of conducting intimate lives. For Giddens (1991), as individuals participate in the labour force, one's life choices, pathways, aspirations, and arrangements have become individualised. Notably, in contemporary times, the emancipation from existing social structures that define partnerships and familial arrangements (Bauman, 2000) has been mobilised by the advent of ubiquitous digital communication technologies and mobile dating applications. Individuals no longer meet their partners within a locality. They are also ushered into what Constable (2005) refers to as "global hypergamy"

Soriano, Cheryll Ruth and Earvin Charles Cabalquinto: *Philippine Digital Cultures: Brokerage Dynamics on YouTube*. Amsterdam: Amsterdam University Press, 2022
DOI: 10.5117/9789463722445_CH04

or union of individuals coming from diverse backgrounds living across countries. The interaction, union, and eventual settlement of proximate or geographically dispersed individuals are facilitated by either cross-border mobility or virtual forms of intimate exchanges (Constable, 2003).

This chapter focuses on the brokerage of ideal interracial relationships between Filipino women and their foreign male partners on YouTube. It examines ten videos produced by Filipina YouTubers, with their contents showcasing the collation of personalised stories around their intimate relations with their partners. A critical examination of these selected contents problematises how the aesthetics, rhetoric and platform-specific strategies of gendered, racialised, and classed representations broker, perpetuate as well as challenge pre-existing representations of an intimate interracial partnerships involving Filipinas. To approach this, we connect our examination of the narratives to the broader socio-historically specific contexts through which representations, meanings and mediated practices are produced. In the Philippine context, the ascribed and discriminatory meanings of interracial relationships are informed by the historicity of the uptake of web technologies among Filipinos who aspire for, engage and end up in interracial relationships (Constable, 2003). For instance, scholars have pinpointed that the emergence of matchmaking technologies has been reported and known to facilitate the flows of mail-order brides in the Philippines, highlighting how women from poor economic backgrounds are just after a foreign visa and money (Constable, 2003, 2006; Rondina 2004). Through the historicity of partnerships and marriages between a mail-order bride and a foreign man as facilitated by web platforms, the figure of the Filipina has been consistently associated with mail-order brides (Gonzalez & Rodriguez, 2003). However, scholars have argued that such reductionist representations often elide the diversity of factors why a Filipino woman would end up in an interracial relationship (Constable, 2003; Saroca, 2002).

Several scholars have unveiled the politics behind the desires for interracial partnerships. They have argued that Filipino women are commodified in matchmaking platforms, often placing them in online catalogues and inviting the gaze of men who can "afford" to purchase a desired woman (Rondina, 2004; Tolentino, 1996). They have also pointed out that the capacity and desires for having a Caucasian partner are tied to the Philippines' colonial past (Tolentino, 1996). Through exposure to cultures, traditions and language of the colonisers, the Filipino's taste, views of the world, and understanding of social mobility have been contaminated with colonial meanings (Tolentino, 1996). The desire for partnering, establishing affinities, or being associated with the colonisers stems from aspiring for the privilege that has been ascribed to foreigners and even fair-skinned people (Rafael, 2000). As presented in Chapter 2, the desire for a Caucasian partner is associated with raising one's social status by having a mixed-race child (Hunter, 2005; Glenn, 2009), which symbolises intergenerational mobility and privilege especially when "whiteness" is ascribed with positive connotations (Rondilla, 2009). As such, to date, there is a salient number of interracial partnerships between Filipino spouses and foreign nationals, who are not necessarily Caucasian. For instance, with the growing number of Koreanovelas in the country, there are Filipino women who moved to Korea to marry Korean men through "marriage brokers" (Garcia, 2011). Nonetheless, although the source of matchmaking is not particularly identified, according to the statistics released by the Commission on Overseas Filipinos, there were 559, 944 marriages between Filipino spouses and foreign nationals from 1989–2018 (Commission on Filipino Overseas, 2018). Over half a million, or 511,453 Filipino women and 48,491 Filipino men were married to a foreign national from 1989 to 2018 (Commission on Filipino Overseas, 2018). The top three countries with the most number of foreign nationals with Filipino spouses from 1989 to 2018 are the U.S. (245,219), Japan (125,813) and Australia (42,644) (Commission on Filipino Overseas, 2018).

In examining the selected videos of YouTubers who curate and visibilise their interracial relationship and broker idealised interracial coupling, we particularly identify "LDR" or a colloquial Filipino term for "long distance relationship" as a central theme of the videos. The word "LDR" was a result of our keyword search of the words "Filipino and Foreigner" in Google trends. For this chapter, we situate our examination of the LDR-themed videos through the lens of countererotics or the utilisation of the Internet to counter or contest sexualised, objectified and stereotypical representations of individuals in interracial relationship (Constable, 2012). Here, we highlight how creative and personalised representations of intimate relations by YouTubers indicate agentic practices (Constable, 2003). Importantly, we further examine online performances as informed by neoliberal and postcolonial contexts, noting how postcolonial performativity is showcased in diverse texts (Spivak, 1999). However, we also argue that the aesthetic and lexicon of enacting intimate interracial relationships adheres to the process of what Arnado (2019) conceptualises as "cultural whitening," which conception was deployed in Chapter 3. Through cultural whitening, the partnership or marriage of a Filipina to a white man indicates the residues of colonial desires. An interracial partnership – in particular a marriage to a white man – is perceived to facilitate social mobility and a stable future

(Rondina, 2014). Its attainment therefore aligns with a positive ending in a "fairy-tale" like quest.

We also analyse the videos as part of digital and commercial cultures, highlighting our contention on how interracial intimacies are brokered. We argue that the portraiture of cultural whitening is co-opted for profit making among YouTubers. The contents broker imaginaries of happiness and the promise of a good and intimate life among viewers. Further, gendered and racialised ideals are reinforced through visualisation, curation, and storytelling. Through these key points, we posit that the visibility and performativity of Filipina YouTubers as symptomatic of postfeminist entrepreneurialism, signalling how narratives and aspirations of interracial marriage can be utilised for entrepreneurial practices and branding strategies in digital environments (Banet-Weiser, 2012). This means that YouTubers broker the idea that mobility, intimacy, and happiness can be achieved through the consumption of digital channels. This manifests in how Filipinas discuss their personal experiences in the online dating scene and sustaining an LDR arrangement as the relationship progresses. Moreover, the "quest" for finding the right one is also utilised to attract followers, generate views, and earn profit on YouTube. Complementing the presentation of valorised success in matchmaking practices is the operations of algorithms, placing contents in a cycle based on popularity and similarity (van Dijck, 2013). Ultimately, the curation of interracial and intimate encounters on YouTube offers a vantage point to re-think the brokering of intimate relations in a postcolonial, neoliberal, and digitalised environment. Before we discuss the findings of the chapter, we map how the socio-historical, socio-economic, socio-political, and socio-technological contexts have shaped the notions, practices, and meanings of an interracial relationship in the Philippines.

The Construction of Postcolonial Intimacies

Several studies have discussed how the logics of desires for an interracial relationship among Filipinos are shaped by historical, political, and economic forces. Filipino scholar Vicente Rafael (2000) argues that racial hierarchy, as a form of colonial legacy in Philippine society, influenced the attitudes and practices of Filipinos towards Westerners and mestisos/as. He articulates that the social hierarchy that began in Spanish colonisation has instilled in the Filipinos' mind the privilege and dominant positions of the colonisers and the mestisos/as. As discussed in Chapter 3, the Spanish colonial period ascribed fair-skinned people with imagery of civility, beauty, and success (Rafael, 2000; Rondilla & Spickard, 2007). Moreover, mestisos/as or mixed-race heritage (typically individuals with a combined indigenous and Spanish and Chinese background) often held a privileged position in political and economic sectors (Cabañes, 2019; Rafael 2000). In contrast, indigenous Filipinos were associated with barbarism, violence, and deviance, which then contributed to legitimising control and policing by the colonisers (Rafael, 2000). Moreover, as part of a colonial legacy, Filipinos position Westerners on the top of the racial ladder, followed by Orientals such as Japanese, Koreans and Chinese, and Indians (associated to loan sharks), Middle Eastern (associated to terrorism), and African people (Cabañes, 2014, 2019).

The high regard to foreign nationals continued during the American period. As presented in Chapter 1, the American period saw the deployment of benevolent assimilation. This approach allowed the colonisers to frame "American rule" as friendly and akin to familial relations. The result was the provision of a range of support systems among Filipinos, such as placing elite Filipinos in government positions, the distribution of American-based media contents, and the implementation of a US-patterned education system (Aguilar, 2014). These key ingredients in the process of benevolent assimilation paved the way for the exposure, appreciation, and further colonisation of Philippine society. The portrayal of American cultures, traditions, and practices have defined standards of a good, secure, and advanced life (Tolentino, 1996). Moreover, the mobility and settlement of several Filipinos in the U.S. mediate colonial desires. Narratives on the good life of nurses in the U.S., and the success stories of those who studied in the U.S. and secured a high position in Philippine society, constructed the colonialist project as legitimate and desirable (Aguilar, 2014; Choy, 2003). Nevertheless, colonial influences are powerful frames that impact cultural taste, aspirations, and practices among Filipinos.

In the context of interracial marriages, Tolentino (1996) argues that the logic of desire to search and marry a foreign partner is informed by colonial, capitalist, and militarist structures in Philippine society. Tolentino (1996) points out how the presence of U.S. military camps in the Philippines and the work opportunities available for Filipinos in the U.S. reconfigured intercultural sociality and intimate relations (Tolentino, 1996). For the former, Filipino women, and also children, were subject to prostitution as military camps populate some regions in the Philippines (Tolentino 1996). The lack of job opportunities in some areas in the Philippines compelled some women to opt for sex work and prostitution. By working as waitresses and entertainers in military camps, women met foreign men, which led to partnerships, marriages, pregnancies, and unfortunately for some, abuse. As such, the word *amboy* became popular, which meanings are ascribed to a mixed-race child or particularly a child of an American man and a Filipina. The hospitality jobs of Filipino women were also evident in the number of Filipina nurses who assist American soldiers in base camps (Tolentino, 1996). Nonetheless, these are some work trajectories of Filipino women during the American rule.

The Philippines remains a neo-colonial state, which stirs the enactment of interracial encounters and relationships. Although the U.S. granted the Philippines its independence in 1945, the Philippines has remained a neo-colonial state because of the nation-state's economic dependence on the U.S. through neoliberal policies (Tolentino, 1996; Aguilar, 2014). U.S. military bases remain in the Philippines and the Philippine economy has been operationalised to the demands and policies of the International Monetary Fund (IMF) and World Bank. The Philippines borrowed money from such institutions, and in return, it opened its economy to deregulation, privatisation, and liberalisation. This political and economic landscape has then contributed to racialised, gendered and classed structures that exploit Filipinos. This is reflected in the exportation of migrant Filipinos (Rodriguez, 2010; Guevarra, 2010) as well as the commodification of women's body (Tadiar, 2004) through prostitution and sex work (Tolentino 1996; Roces, 2009). Given the affiliations developed during and after American rule, the advent and use of letters, catalogues, and eventually web technologies have contributed to the continuities of harnessing interracial partnerships. However, the Filipino woman's body became deeply associated with the term "mail-order bride." Here, a Filipino women's marriage to a foreigner is understood to be a form of neo-colonial fantasy or the fulfilment of inter-racial ties that carry the promise of a good life (Tolentino, 1996). However, for the male foreigner, marrying a Filipina fulfils a nuclear family fantasy, positioning them as dominant in the relationship while the woman performs a submissive, domesticated, and nurturing role (Tolentino, 1996). Notably, some interracial marriages come with diverse issues. Some women fall as victims of human trafficking, sex slavery, and domestic abuse (see Garcia, 2011; Kusel, 2014; Lee, 1998; Lindee, 2007). In contrast, Filipino women are also portrayed as abusive in a relationship or someone who will dupe men (Parreñas, 2011). Moreover, the violence against women is also legitimised by framing women as fraud or individuals who are just after a marriage with a foreigner to obtain permanent residency (Amy, 1997).

Other scholars have argued to locate a sense of agency of Filipinos who meet and marry foreign nationals in various matchmaking channels by moving away from simply engaging in representational politics (Constable, 2003; Parreñas, 2011; Saroca, 2002). For instance, by conducting an ethno-graphic study of Chinese and Filipino women who married U.S. nationals through matchmaking channels, Constable (2003) has argued how Filipino women demonstrate their agency in various ways. First, not all women who access websites or engage with matchmaking agents consider themselves as being commodified and sexualised. Constable highlights that markers of difference, such as class or educational background, tend to contribute to the agency of women in handling interracial relations. Further, Filipino women have different motivations, intents, practices, and strategies in finding a romantic partner and being in an interracial relationship. For example, some Filipinas opt to marry a foreign national because they could not find a local man who can accept their cosmopolitan ideals of romance (Cabañes & Collantes, 2020), or their living conditions such as being widowed, a single mother, or too old for a local man's taste. Additionally, the choice of images, text narratives, as well as communication process, allows for the formation of self-actualisation (Constable, 2003).

More recently, Constable (2012) has also highlighted how Caucasian men who marry Filipino women call out discrimination and stereotypical representations of Asian women through virtual communities, such as newsgroups and forums. Countererotics, as a concept, pinpoints how Filipino women veer away from sexualised and exoticised representations. Here, Constable (2012) coins the term countererotics to describe the oppositional narratives and testaments of those in interracial relationships against the sexualisation and objectification of women. Despite such contention, Constable (2003, 2012) argues that emphasising the agency of women in interracial relationships should not be romanticised. Women may face discrimination from the family of their foreign husbands (Constable, 2003; Parreñas, 2011) or even in their ethnic communities (Aquino, 2018). They can also be subject to unwanted requests from their family members because of perceiving their marriage to earn and send dollars (Constable, 2003). Nonetheless, by presenting the challenges that women bear and negotiate in interracial relationship, abuse and exploitation are not elided (Constable, 2003).

It is important to note that women's agency in postcolonial conditions are reflective of the social structures women navigate in Philippine society. In this regard, we engage with the work of Arnado (2019) on cultural whitening. By examining the lives of Filipino women married to white men, Arnado (2019) coins the term cultural whitening to capture and explain the desire for whiteness among Filipino women who marry white men. Specifically, she

proposes three states of cultural whitening: (1) embodied (habits), (2) objectified (consuming goods and products), and (3) institutionalised (including
education, migration, marriage, and citizenship). For her, the desire for
whiteness among Filipino women stems from the opportunities enabled by
"being white" – such as accessing economic mobility, physical security, and
assimilation in a multicultural country. This connects to Tolentino's (1996)
notion of "neo-colonial fantasy" as reflected in the desires for an interracial
partnership or marriage. However, Arnado (2019) highlights that Filipino
women feel ambivalent towards their transformed lives. Here, the more
they become "white," the more they lose their connections to their roots.
The whiteness is evident in living overseas, a reconfigured sense of decision
making, speaking a foreign language, and so forth. As such, they typically
employ forms of resistance such as continuously practising Filipino traditions and upbringing via cooking or retaining their Philippine citizenship.
However, as Arnado (2019) notes, cultural whitening is an indication of
how Filipino women navigate gendered, racial, and even classed hierarchy
enabled by colonial legacy in the Philippines.

As we have presented, interracial relationships in the Philippines are
informed by socio-economic, socio-political, and socio-historical conditions.
However, it is important to note that those in interracial relationships also
demonstrate agency through diverse practices, despite the contradictions
underscoring this exercise of agency. In the next section, we turn our discussion to the role of fast-evolving media channels in shaping the enactment of
interracial ties and relations. Technological transformations are crucial in
shaping the agency and experiences of Filipino women, which are leveraged
on for brokering an interracial and intimate relationship.

Continuities, Contestations, and Contradictions in Mediated Worlds

Even before the advent of dating applications, various matchmaking tools,
including video tapes and letters (Constable, 2003), catalogues (Tolentino,
1996), websites (Del Vecchio, 2007; Zug, 2016) and matchmaking agencies
(Constable, 2003) have been utilised by men and women who are in search
for a romantic partner. In the Philippines, these matchmaking outlets have
been analysed through textual and critical discourse analysis, highlighting
how women are typically sexualised and commodified (Tolentino, 1996)
as well as exoticised and eroticised (Robinson, 1996; Starr & Adams, 2016).
Meanwhile, white men are framed as customers of these women, positioning

them as strong, dominant, and "saviours" of women (Tolentino, 1996; Zug, 2016). Nevertheless, the dichotomous representation of gendered, racialised and classed individuals often depicts power differentials in a "transactional" or "economised" modality.

Matchmaking channels have diversified over the past years. Certainly, web technologies still exist and facilitate interactions, meet-ups, and eventual marriage (Rondina, 2004). More recently, a diverse range of social media channels and dating applications have surfaced, providing new ways of enabling interactions and intimate relations. These dating applications can be argued as a continuity of previous matchmaking channels that allow individuals to present themselves through profile photos, texts, and other digital information (Ellison, Heino, & Gibbs, 2006). These channels have also facilitated a space for non-heteronormative relationships (Licoppe, 2014; Hobbs, Owen & Gerber, 2017; Quiroz, 2013). In the context of the Philippines, dating applications are becoming popular. According to the report released by YouGov Philippines in 2017, half of the 2,777 surveyed Filipinos used an online dating app (YouGov Philippines, 2017). Further, recent studies show that digital matchmaking channels allow Filipinos to meet foreign nationals as intimate partners and embody a sense of cosmopolitanism (Cabañes & Collantes, 2020). However, discrimination arises from existing prejudices to certain racialised, gendered and classed bodies in interracial matchmaking. Here, interactions and relationships are perceived as transactional, and Saroca (2007) notes as "akin to prostitution and devoid of romantic love" (p. 82).

In this chapter, we highlight the representations of intimate and inter-racial relationships on YouTube. This is of great importance as networked and personalised media are becoming critical spaces to display, affirm, and even counter racialised, gendered, and classed relations (Constable, 2012). In these spaces, personalised and localised contents (Hjorth, 2011) and vernacular creativity (Burgess, 2007) become focal points for participating in an intimate public (Dobson, Robards, & Carah, 2018). On YouTube in particular, ordinary individuals curate their personal, quotidian and intimate experiences (Strangelove, 2010), which trigger opportunities for interactions and connections in an online community (Burgess & Green, 2018; Lange, 2007). In an online space though, performativity is reflected in how people present themselves, which indicates strategies to navigate the public and private self (Lange, 2007). As we will show in this chapter, the YouTubers broadcast their intimate relationships to either affirm or negate notions of an interracial arrangement. The exposition of intimate, challenging and success stories function as important sources for the brokering of interracial intimacies.

An online space like YouTube is not a neutral space (van Dijck, 2013). As discussed in Chapter 1, it has its own business model and governance which can shape the performativity, content circulation, and other forms of mobility in digital environments. Examining YouTube allows us to reflect on how the blurring of boundaries between amateur and professional contents has been made complex especially when new systems and processes impact who and what becomes visible, popular, and profitable. For instance, the emergence of multi-channels networks on YouTube shapes the privileging of contents among content producers who are not necessarily amateurs and not also professionals (Lobato, 2016). At the same time, ordinary or amateur producers tend to negate YouTube's privileging of celebrities' visibility in the online space, noting how YouTube has initially positioned itself as a space for community making (Burgess & Green, 2018). Evidently, what this presents is that the governance of the operations of YouTube is primarily controlled by its business model. Profit-making is achieved through which contents adhere to a popularity principle – the more likes, shares, and comments then the more chances of having contents embedded with ads. With this understanding, content producers create their own and unique strategies to standout in a competitive and saturated market (Marwick, 2013). Here, mundane, intimate and relatable contents are capitalised on by YouTubers or Internet celebrities to generate value and profit (Abidin, 2018, García-Rapp, 2017; Glatt & Banet-Weiser, 2021).

In the context of interracial relationships, for instance, by examining how interracial relationships are constructed on YouTube, Sobande (2019) argues that intimate interactions are visibilised and romanticised online to garner attention and views among audiences. Further, online narratives, constructed to appeal to demands of audiences that fetishise interracial relationship, may reinforce heteronormativity and racial objectification (Sobande, 2019). Building on this point, the online space potentially demonstrates a form of postfeminist branding (Banet-Weiser, 2012). For Banet-Weiser (2011), YouTube has been utilised by women as spaces for empowerment – visibilising themselves and generating a collective support. Online spaces facilitate exchanges that mediate a sense of belonging and community. In contrast, digital practices can also become a locus for tension especially when generating likes, views and shares depends on the audiences that have internalised social structures. Here, visibility must conform to audience expectations, which the content creator has to manage through various strategies of relational labour (Baym, 2015). This point echoes Sobande's (2019) study, which showcased how women in interracial relationships tell a particular narrative that will entice the viewer's expectations, including reiterating heteronormativity, romanticising interracial relations, and even fetishising a mixed-race baby. Overall, online spaces become sites of contradiction especially when social structures, governance, and business models shape the contours of intimate ties. We now discuss the videos of selected YouTubers who visibilise their interracial relationships on YouTube.

Intimate Expressions on YouTube

This chapter examines ten videos produced by ten female YouTubers from the Philippines who are in an interracial partnership. These videos were collected using Google trends. Initially, we used the keywords "Filipino and Foreigner" to identify the most popular words used in the Philippines to search for YouTube videos on interracial partnership. The keyword "LDR" emerged, and this word was then used to access, filter and analyse YouTube videos about the intimate partnership between a Filipino and a non-Filipino. In July 2020, 100 videos were collected and viewed based on the results via YouTube search. Out of the 100 videos, we selected the first 50 videos with the highest number of views and subscribers. From the 50 videos, we selected 10 videos and transcribed the contents. The study shows that the YouTube videos of 10 Filipina YouTubers (refer to Table 4.1) primarily showcase the activities involved in the formation and maintenance of a long-distance relationship. In our examination of the selected videos, we highlight how the contents of each video act as mechanisms for the brokering of intimate and ideal interracial partnerships. We then categorised the themes of the videos in relation to our conception of brokering on YouTube, including affective content, discursive style, credibility building and platform-specific strategies.

   We observed ethical considerations in our data collection, analysis, and presentation, following the guideline of Deakin University's Human Research Ethics Committee (DUHREC) (project number 2018–364). Similar to Chapter 3, we also followed the ethical decision-making and recommendations of the Association of Internet Researchers' Ethics (2012, 2019) by ensuring an ethical handling of selected and publicly accessible videos. This chapter employed anonymity to protect the privacy of individuals in interracial relationships. We de-identified the name of the YouTubers and their partners by using a pseudonym. We also used a generic title for the videos. We also removed identifiers, including country of origin, settlement, and other personal information. The quotes incorporated in this chapter

Table 4.1   Relationship brokers and their videos curating interracial relations

YOUTUBER	JOINED YOUTUBE	NUMBER OF SUBSCRIBERS	NUMBER OF VIDEOS	VIDEO ANALYSED	DATE PUBLISHED	NUMBER OF VIEWS
Jasmine Cruz	3 March 2012	1.11 M	1,581	Our love story	26 June 2016	1,190,945 views
Sheree Dalisay	18 July 2017	25,100	12	Long Distance Relationship Meeting For The First Time!	17 August 2018	3,568,189
Cathy & Dayne	7 Nov 2013	26.3	37	Long distance relationship story	30 September 2017	1,921,442
Team Power Hugs	3 November 2015	10.5K	46	LDR meeting for the first time	7 May 2016	1,269,315
Triple Hearts	20 December 2015	20,000	242	LDR!!! Best surprise visit ever – couple reunited	1 May 2016	811,424
Lovers Lane	22 Aug 2016	3.6	24	Our LDR story to marriage	3 Feb 2018	332,507
Sonny & Honey For Life	22 Oct 2009	1.82	56	Long Distance Relationship and meeting for the first time	4 May 2017	164, 971
Happy Place	21 Dec 2010	4.73	74	My Korean partner and our LDR journey	14 Jun 2020	88,167
Marcus and Anna's Journey	15 December 2018	6.25K	50	LDR no more!	25 April 2020	85,926
Cathy Lim	4 September 2013	14,600	175	Tips on keeping your partner in a long-distance relationship	20 July 2020	68,408

were based on the contextual translation of the original texts (Zimmer, 2010). These approaches allowed us to protect the privacy of the YouTubers.

As we present in the following sections, the YouTube videos can be considered countererotics as they challenge the sexualised and objectified imagery of a Filipina in an interracial relationship (Constable, 2012). However, despite the attempt of representations to subvert the sexualisation of women, the aesthetics and rhetoric of the videos tend to convey the process of cultural whitening, which reinforce hegemonic social structures (Arnado, 2019) especially in a commercial and gendered (Banet-Weiser, 2011) or racialised digital environments (Sobande, 2019).

The tale of finding a foreign partner

A distinct feature of the YouTube videos is the affective and personal storytelling of enacting a long-distance relationship through an ordinary person's voice and lenses. It is common for the videos to start with the story of how a Filipina or the Filipina YouTuber met their foreign partner. Here, the YouTuber articulates what online platforms were used that led to online interactions. The YouTuber then shares how more intimate, personalised, and creative interactions are moved and developed in another platform, such as Skype, Facebook or Viber. These practices show that the effort to meet and sustain an intimate interaction occurs in an environment populated by multiple devices and online channels (Madianou & Miller, 2012). The visuals and texts are utilised through a messaging application or longer conversations through videoconferencing. After presenting the exchanges in different platforms, the narrative is followed by the physical meetup of the Filipina and the foreign national. Often, this moment is captured in the airport, showcasing the arrival of the foreign national and being met by the anxiously waiting Filipina. Intimate moments of being physically together are captured. Eventually, the "reunited" couple are typically seen travelling together and even enjoying several tourist spots in the Philippines. Further, the foreign national engages in a "meet-and-greet" session with some family members and friends of the Filipina. However, the videos also show that the foreign national had to leave again and an LDR arrangement is opted to sustain the relationship. Dramatic moments in the airport are presented, showing the couple saying goodbye to each other. As such, the moment of separation and shifting back to a mediated or long distance relationship may then entice the viewers to continuously follow the YouTuber's quest for overcoming the challenges of a distant coupling. As if following the plot of a regular television romantic series one is left with a question, will the relationship work? The success of an LDR is then gauged when a relationship lasts or ends in a marriage and physical reunification of the couple.

A case in point is the video of Marcus and Ann's Journey, showing a fairy-tale-like love story. In the video, viewers are informed that after seven months of online communication, particularly video call, Marcus and Anna decided to meet in the Philippines. They are shown to be having fun, cuddling, dancing together, and even touring places. Eventually, Marcus asked Anna to marry him, a union to be sealed upon the blessing of Anna's parents. Marcus returned overseas and they went back to online communication. After seven months, Marcus came back to the Philippines. They got married. The video ends with a statement from Anna about preparing for a visa application. The happy ending of the story is captured through the marriage photos.

It is without a doubt that the narrative presents a modern-day fairy tale. A woman meets a man through an online channel and eventually ends up in a marriage. Over the past years, this aspirational trajectory has been capitalised by various dating platforms that market matchmaking opportunities for individuals who are in search of a local or overseas partner (Constable, 2003; Tolentino, 1996; Zug, 2016). Supporting the saleability of a matchmaking process is the inclusion of testaments from "satisfied" or "happy" customers, who are essentially able to find and marry their dream partner through an online site (Constable, 2003, 2012). However, in these web technologies, the main feature of matchmaking focuses on the "meetups" and the happy union or marriage of two individuals. What is missing though are the tiny and most intimate details of meet ups, the courting stage, and activities leading up to a marriage. This void, we argue, is now filled by the narratives of Filipina YouTubers who share their everyday experiences and travails in finding their partners online. Elsewhere, we have highlighted how a Filipino woman in an interracial relationship showcases the intimate details – ups and downs – of sustaining an interracial partnership (Cabalquinto & Soriano, 2020). In our study, the YouTubers used their own experiences to showcase the challenges and benefits of meeting an online partner while also developing a persona for themselves and their partner to engage audiences and eventually monetise their contents in an online platform such as YouTube. They present the actual steps in enacting interracial encounters, and more importantly, the triumphant feeling when a long-distance relationship turns into physical reunification. In a sense, the "investments" put into meeting the partner online, and sustaining relations beyond borders that eventually lead to a marriage or long-term partnership (Constable, 2009), while generating profit from YouTubing indicate the commodification of intimacy.

A case in point is the YouTuber Jasmine Cruz and her video entitled "Our love story: Storytime." During the conduct of this study, her video had a total of more than one million views. Through a storytime genre or a confessional format (Raun, 2018), her video details how she met her Caucasian partner through a dating website. Her video narrates how she signed up for the online channel to speak to someone but not really after "having a boyfriend." She confesses that she just broke up with her boyfriend and wanted to take things easy. She also highlights, on the dating website, she met her partner, Jack. Back then, her partner responded to her through the dating website and they eventually shifted their exchanges on Viber. They chatted for six months through a combination of various online channels. Eventually, they decided to meet in Malaysia, an encounter convincing her of Jack's kindness. Eventually, after completing the trip to Malaysia, they decided to visit the Philippines, which led to Jack meeting her parents. Eventually, they became a couple and sustained their long-distance relationship. Jasmine, after several months of staying and waiting in the Philippines, followed the man overseas. She visited him with a tourist visa but eventually went home again and then back overseas when a de-facto visa was released. Ultimately, Jasmine's journey exemplifies a success story.

However, despite the success story, Jasmine warned her viewers about signing up on online dating websites. She said, "To all my viewers who are 18 years and below, please take care. I won't tell you not to sign up, it's your life. But please take care. It can be dangerous." This statement presents the anxiety that Jasmine felt when she was about to leave the country and meet Jack, but also signals the YouTuber's acknowledgement that her stories can influence her viewers' actions. She mentioned that she was feeling anxious about a meet-up. Yet, she also highlighted that the "constant communication" with Jack pacified her fears. For her, the consistency of exchanging messages, photos, and even videoconferencing act as evidence of Jack's genuine intention. As Jasmine had unsettling experiences online, such as being messaged by random men wanting to marry her, she reminded her viewers about the importance of being cautious in an online dating scene. Nonetheless, YouTube provides a space where women discuss individual safety by highlighting the important role of "getting to know" someone through constant communication.

YouTubers construct their persona and their partner's persona on YouTube. In the first instance, the YouTubers present themselves as women who are "willing to wait." They showcase this in different dimensions. First, they present the virtue of "patience" by engaging in the use of online channels to forge and maintain interactions with their partners without aggressively demanding for an immediate meetup or marriage. Second, they perform such acts right after the partner goes back overseas. In such performativity, what is presented is how a person's "perseverance" gets to be rewarded in the end. The waiting period becomes a testing period for knowing each other and gauging if the relationship will work. Notably, the culmination of enduring a long distance or technologically mediated relationship is transformed into a partnership or marriage.

An evident example of this is the music video of Cathy and Dayne, "Long distance relationship story 2017." In their video, Cathy talks about the challenges of living apart. She shares that Dayne left the Philippines in 2016 and moved back overseas. She exclaims, "I was feeling lonely when he left, we got used to being next to each other already. After one day, when he got back to his city. He called me and I was so happy but still sad. We were so emotional and couldn't help but cry." This statement is then followed by a narration of how they sustained the relationship using different online communication channels. Special occasions celebrated through digital media use are shown. For instance, during Cathy's birthday, Dayne called her on video, but also sent some surprise gifts. Meanwhile, during Dayne's birthday, Cathy prepared a cake, and Dayne was on video call. In a sense, these practices of sustaining intimate ties at a distance demonstrate how both individuals are enduring the long-distance relationship through online communication and other forms of emotionally charged and ritualised activities such as sending gifts and celebrating from afar. In a way, Cathy's ability to wait and engage in long distance communication demonstrates the characteristics of a loving and caring partner. Her videos prove that a long-distance relationship with a foreign partner can work with the right character and effort, and this can then lead to eventual reunification. As she highlighted, "LDR is NOT easy and it will never be. When the person you love the most is far from you. All you can do is to wait 'til you meet again. But you will wait coz you know it'll be worth it."

Women in the videos also demonstrate their agency in an interracial relationship. They present their decision-making abilities before going into a relationship. For example, in the video of *Happy Place*, Regine and her Korean husband Teo recalled the day they met. Interestingly, they did not meet through a matchmaking website. Regine detailed that she was just waiting for the results of her nursing board exam in 2008 when she decided to look for a part-time job. She ended up teaching in an English Academy, which specialises in teaching English to Koreans in the Philippines. At the time she was teaching in the academy, her now husband became her student. Fast forward, they became official partners two weeks before Teo left the Philippines to go back to Korea. That moment began their seven years of long-distance relationship. Like other YouTubers in this study, the foreign national typically visited the Philippines after a few months of being away from their Filipina partners. In the case of Regine, Teo returned to the Philippines two months after he left the Philippines, also at a time when Regine's mother was celebrating its birthday. Interestingly, Teo's parents also visited the Philippines in September 2010, which Regine highlighted, "Koreans actually introduce their special someone (may it be their girlfriend/ boyfriend or fiancé) to their parents or relatives once they're sure that they'll marry that person."

Although this background story of their marriage was mentioned, compared to other videos that we examined, Regine did not reveal the details of how they maintained the long-distance relationship. What she highlighted though is her migration to Korea on a student visa, showcasing the effort to invest in a potential partnership. As she said, "So I came here in 2015 but I didn't come here on a marriage visa because we're not married. I came with a student visa. My reason for that is that I told Teo that most of our relationship we spent apart so we don't know if we're really compatible. Because even it entailed a lot of money because you have to enrol to at least two semester coz that will be the duration for your visa so I took that route instead of getting married." Eventually, they got married in 2017.

The character of the foreign national is also constructed in the videos. Often, the foreign national is presented as a kind, welcoming, and genuine person. This approach contributes to the brokerage of an ideal interracial relationship. It highlights the importance of "choosing" the persona of a foreign partner that can harness a good and meaningful relationship. As such, YouTubers showcase their partner's characteristics through curated activities.

First, the YouTuber points out the unique and endearing characteristics of the man, which paves the way for the development of special and intimate feelings. For instance, Jasmine talked about how her feelings for Jack developed.

He says that I am "chaka" (ugly) in my photos in the dating website so I eventually like him. I like people who are like that. Again, to be honest, there were so many guys in that website that even though I put my ugly photos they will still tell me, "Oh my god, you're so pretty. I want you to be my wife." I am like, what? The reason why I put my wacky photos in

there because I don't want people to visit my site because they can see my
cleavage in my picture. I want them to visit my site because they want to
speak to me, and they are serious to find a special someone. To be honest,
I really don't have time to chat or flirt with guys.

In the above statement, the foreign national has to pass a "test" based on
kindness. But more importantly, it counters the existing and generalised
assumptions about matchmaking sites where overseas men find and buy
a bride.

Second, the effort of an overseas man to travel to the Philippines despite
the vast distance was used to frame a man's authenticity. In most videos,
most physical meetups happened in the Philippines after a routinised online
communication. The Filipina is often the one who waits for the man. In a way,
the visit of an overseas man becomes a symbolic conquest, demonstrating
the effort, dedication, and commitment to pursue the relationship. For
example, in the video of *Triple Hearts*, Patrick, who is based in Australia
was planning a surprise visit to the Philippines. At the outset of the video,
he is seen chatting with his partner, Bonnie. He mutes the microphone in
his mobile phone and walks towards the woman. On screen, the viewers
can hear him saying to Bonnie that he is visiting the Philippines. But with a
muted microphone, Bonnie couldn't hear it. The phone conversation ends,
and Patrick begins his long journey to the Philippines for a surprise visit.
His journey is documented from the airport up to landing in Manila and
reaching the area where Bonnie resides. He is shown approaching some
locals to ask where Bonnie lives. In one encounter with a man on the street,
he shows Bonnie's photo and asks the man if he knew her. Luckily, the man
recognises Bonnie and then accompanies him to Bonnie's place. Bonnie
sees him and she looks very surprised and happy. In this video, viewers
are delighted with the conquest of the man, showing the commitment in
the relationship.

Lastly, the foreign national is portrayed as adaptable. This means that the
man enjoys Filipino food, meets the woman's friends and family members,
and learns to adjust in a new environment, even for a short period of time.
For instance, going back to the video of *Triple Hearts*, Patrick is seen not
only overcoming the challenges of travelling to the Philippines, as shown
in how he walks in narrow pathways in Manila. But he is also presented as
showing respect to Bonnie's family by doing Filipino practice of respecting
the elderly – the *mano* or a Filipino gesture of respect wherein an individual
bows and presses one's forehead on the back of the hand of an older person.
Similarly, the video of Marcus and Anna's Journey shows how the man

respects the woman's family. In the video, Marcus asks for the blessings of Anna's parents. Through a photo embedded in the video, Marcus engages Anna's father in a serious conversation. And in the video of Jasmine Cruz, Jack is described as someone who is "patient" during his stay in the Philippines. She says, "You can tell if your foreigner boyfriend is patient when he showers without a heater. We don't have a heater at home. The water coming out of the shower is cold. But he survived. He took a bath with a cold shower!"

YouTubers presented content that visibilise the story of meet ups, use of digital technologies to sustain a long-distance relationship, and the eventual reunion, marriage, and settlement elsewhere. This narrative structure is presented like a fairy-tale where a woman and a man find each other and live happily ever after (Rondina, 2004). However, instead of highlighting the "outcome" of exchanges in online spaces, the YouTubers construct their persona and their partners as "authentic" and "committed" in the relationship. This approach convinces the viewers that getting involved in an interracial relationship is more than using a digital technology and enabling connections. For the YouTubers, forging and maintaining interracial intimacies necessitates a critical investment of time, feelings, and constant communication, reiterating the "genuine" aspect of a relationship or sincerity of the couple. We then argue that these contents are constitutive of brokering. The YouTubers, their visibilised stories, and intimate expressions primarily broker the possibility of a "successful interracial relationship" that brings positive affective experiences and potentially, social, and economic mobility. Functioning like marriage brokers, these YouTubers walk their audience through the process of clinching successful interracial relationships through a series of video narratives, selling aspirations while also illustrating how aspirations can in fact be achieved. Like traditional brokers, they put together an array of crucial information albeit this time told across their video series: the comparative advantage and disadvantage of different dating apps, why foreign men are attractive, how to identify a serious man, the value of patience, how to make one's dating profile appear authentic, how to maintain an LDR, the process of securing a marriage or de facto visa, and so forth. These sets of information that are embedded in their stories and confessions serve as a crucial vantage point for other aspirants to obtain first-hand information from fellow Filipinas who likewise desired and have gone through the experience with demonstrated success. The viewers are presented with engaging narratives that tickle a desire of what could be a good and intimate interracial relationship. However, affective representations that are shaped by colonialist legacies are also witnessed, consumed, and lived vicariously by viewers. In the next section, we present the different discursive styles that the YouTubers employ to curate their sweet encounters.

Staging intimate encounters with texts, stickers, and photos

Most videos had a distinct visual, aural, and textual style in presenting the formation and sustenance of a long-distance interracial relationship, which eventually led to physical reunion and marriage. Among the ten videos that we analysed, seven deployed a music video format. This means that YouTubers told their intimate story through a montage of graphics, photos, and even additional emojis and stickers. Additionally, the music set up the mood in the love story, often representing the emotions that the couple went through in sustaining an interracial relationship or marriage, including an online meet up, sustained online interactions, physical meet up, online interactions, and eventual marriage and reunification. These stages are supported with contents that often reveal the most intimate and personalised experiences of the YouTuber and the foreign partner. Ultimately, in relation to previous discussions, the YouTubers broker the notions and practices of enabling a rewarding interracial relationship.

An example of utilising a music video format to present one's story is the video *Lovers Lane*. The story is told by Lisa from the Philippines, and her LDR journey with Don. The video starts with a fading in music, the song *Perfect* by Ed Sheeran. As this music plays in the background, a series of photos emerge, including the photos of Lisa and Don overlaid with iconic flags. More information is revealed, including the platform they used to sustain their relationship, as well as the timeline and screen capture of online communication prior to the physical meet up in the Philippines. Photos of intimate interactions during their travel in a tourist spot in the Philippines are also shown. Eventually, an image of the engagement is shown, but also followed by a farewell shot in the airport. The video ends with the ceremony of the marriage. The song Perfect plays through the entire video, conveying the sweetness behind the couple's journey and interactions.

Upon close examination, the creative and personalised dimension of the videos are characterised by details that shape the framing of a techno-logically mediated interracial relationship. Evidence of tools, practices and encounters that contribute to the production of genuine and flourishing relationships are showcased. First, the couples' photos are designed with additional stickers and even emojis that showcase the personalisation

Figure 4.1   Still from a YouTube video. YouTubers and their videos broadcasting interracial relations.

of intimate interactions (Hjorth, 2011). In some cases, they incorporate symbols from their respective home countries, indicating their origin. Second, screenshots of online interactions showcase the "authenticity" of the relationship. The personalised exchanges reflect the invested effort and time by both parties in "making the relationship work." This point complements the findings of Saroca (2002) on how Filipino women assess the authenticity of their interaction and relationship building with a foreign man based on consistency and quality of interactions in messaging channels. In this study, however, the YouTubers present a screenshot of video calls and intimate exchanges on Facebook messenger. Third, the YouTube videos broadcast a screenshot of the matchmaking tools as well as other online channels that they used to get to know each other and maintain intimate ties (See Figure 4.1 for an example). Fourth, the videos include dramatic "first meet ups." Often, slow camerawork is used to add a dramatic effect on the meet up. Fifth, travel photos are presented, showcasing how individuals express their feelings and love for each other in a co-located setting after months or years of online communication. And lastly, marriage photos often culminate in the music video format. It shows a marker of success of an interracial partnership that has been built through LDR and also a limited time of physical togetherness during the man's visit in the Philippines. Ultimately, these contents support the framing of an interracial relationship – genuine, worked on, and also safe.

For a confessional set-up, intimate banters and additional props are used. An example of this is the video of *Happy Place*. In the video, the viewers can see Regine seated on a couch with her Korean husband Teo. Regine refers to the set up as "sit-down video." What is interesting though is the presence of big stuffed toys. The set up essentially emulated a cosy living room like that of a talk show. Meanwhile, Regine and Teo are playful and casual in the video. Both greet the viewers with a Korean greeting, *"annyeong haseyo,* guys!" Interestingly, Teo greets the viewers using Tagalog greetings, such as *"magandang araw"* (beautiful morning). Teo also shows his affection to Regine by saying, *"Magandang* Regine" (beautiful Regine), making Regine laugh. The video focuses on the love story of Regine and Teo, overcoming seven years of long-distance relationship before deciding to be reunited and get married. As they share their love story, they are also seen engaging in a funny "stuffed toy" fight. This playful gesture on the video essentially frames the relationship as genuine and filled with happiness. The gesture typically happens when Teo teases Regine on a certain detail of a story. An example is how Regine asks Teo about his past text messages to her when she was working in a hospital in Manila. Regine finds it strange that Teo messaged her given that she already left in the English Academy. In the video, Teo responds to Rona's query on why he messaged her. He says, "Maybe I want to learn English through text messaging!" Both laugh and Regine hits Teo with the stuffed toy. Eventually, the storytelling ends with Regine asking Teo, "Do you think waiting for seven years is worth it?" Teo replies with the words, "Of course!" Regine then exclaims, "It was one of the best decisions of my life. Kinikilig!" with the translation and definition of the word *kinikilig* (elated) superimposed in the video. At the end of the video, the couple does a popular Korean "heart gesture," with the graphics *"Annyeong!"* Ultimately, the set-up, playful banters, funny yet intimate gestures, and additional props and overlay graphics vibilises the past and present of enacting a happy and genuine interracial relationship that has overcome a long-distance arrangement.

We also noticed that some of the YouTubers used labels to brand the relationship. They would use a hashtag that combines their name and their partner's. For example, John and Fiona use "TEAMJF" to label their relationship. The label has been used in the video, such as "After a year of LDR, WE FINALLY MET!!!!#TEAMJF." Significantly, six out of the ten YouTube videos combine the name of the couples as the YouTube channel's name, such as "Cathy & Dayne" and "Marcus and Anna's Journey." We identify such labelling of the channels as a testimony of the depth of the interracial relationship. Importantly, we argue that this approach also follows a distinct trend in Philippine media wherein love teams use a hashtag or code to label their happy relationship. For example, *KathNiel* refers to the combined names of one of the Philippine's most popular celebrity couple, Katherine Bernardo and Daniel Padilla.

Wedding photos are also incorporated in the videos. They often showcase the culmination of all the hardships and sacrifices of the interracial relationship. An example of this is the video of Marcus and Anna's Journey. The couple is shown wearing elegant white outfits as the ceremony took place in the city hall. They are surrounded by family members and friends of the woman, showing their blessing or approval of the union. The officiating judge is also shown, indicating the legality of the marriage. The video ends with the guy kissing the woman, with the image superimposed with the texts "Marcus & Anna." We argue that wedding photos are evidence of a successful long-distance relationship. The marriage represents how long-distance relationships work through consistency and hard work in a duration of months or years of physical separation. Significantly, we argue that visualisation of navigating an interracial relationship counters the stereotypical representation of Filipino women as gullible and passive in ending up in a commodified interracial relationship via web channels.

Curating credibility through the possibility of interracial coupling

Most YouTubers capitalised on their own journey to prove the possibility and positive outcomes of an interracial relationship. They exposed the "pains" and "gains" of navigating an interracial relationship, which is typically tested through a long-distance arrangement. By foregrounding their experiences, they position themselves as having the authority to offer advice and support for viewers who are thinking about venturing in an interracial relationship.

The YouTubers typically showed the hardships of finding a suitable partner, which can also end in a positive and happy outcome. This is the case of Honey and Sonny. In the video, Honey tells how she met Sonny. She shares that Sonny was supposed to meet his chatmate in the Philippines, but it did not happen. Eventually, Sonny went back to the online site. Meanwhile, Honey was heartbroken after her ex, a chatmate, left her. She eventually activated her dating profile, which led her to meeting Sonny. Eventually, they both met online and eventually opted to continue their interactions over Skype. As Honey points out, "Days passed, and we keep in touch. We chat day and night every day until we feel in love with each other." Sonny visited the Philippines in 2017 for a short period, and they went back to a long-distance relationship again. On Skype, Sonny proposed, and Honey said yes. Honey started processing her fiancé visa. With this story, Honey ends the video with the text, "Distance is nothing when two hearts are meant for each other." By closely examining the narrative of the video, Honey establishes her credibility to speak about interracial and long-distance relationship based on her own experiences, "While I am also broken hearted because my ex left me without any reason (Chatmate but never meet in person)."

Experiences translated into "important advice" were also utilised by other YouTubers to establish credibility. For instance, in the video of Cathy and Dayne, fun and challenging experiences of a long-distance relationship are presented, which are then used as a springboard to give advice. As presented in the video, "LDR is NOT easy, and it will never be. When the person you love the most is far from you, all you can do is wait 'til you meet again. But you will wait coz you know it'll be worth it." Similarly, the video of *Team Power Hugs* also articulated this kind of advice, "For those who have LDR out there, don't give up and stay happy. I know LDR is not easy but be patient and when you finally see each other…IT WILL BE WORTH IT!!" In some cases, a YouTuber highlighted the reward of waiting for a special someone. This is communicated by Jasmine in her video, "We are both very happy. We keep on telling each other, I am very lucky, he is very lucky. It's quite good to hear from the man that you love saying that he is lucky that I came into his life. I get teary eyed. But the truth is I am the one who's lucky because I didn't search for him. He just came into my life unexpectedly and my life has become great because of his presence. He makes me so happy. So, love will just come at the right time."

Notably, YouTubers also offered advice on how Filipino women can position themselves in a long-distance relationship. A case in point is Cathy Lim, a Filipina married to a Caucasian man. As per Cathy, the channel works "to share our LDR story and to inspire some couples who are in the same situation." In her video, she gives five tips to women to demonstrate their value in a relationship. First, she points out that women should learn how to handle their anger. She says, "You need to control your anger (*referring to moments of misunderstanding*). Let it pass." Her second tip is about learning to remain calm. In her words, "For us Filipina, it's in our character to be *palengkera* (loud market vendor), *bungangera* (nagger). If you're in an LDR, stop nagging." The third advice is on not being clingy. She supports this advice by pinpointing how one's doubt and mistrust can stir surveillance of others. For her, it is advisable to control one's doubts and make the guy miss you. The fourth advice is focusing on improving oneself. She reiterates that women should make themselves positive, delightful, and easy to be with. She emphasises:

> We often tend to say that looks do not matter for foreigners. Don't believe that. A man will always be a man. If you're fat, and you're in an LDR, then beside him is a sexy woman who likes him, sexy and beautiful, and then you are fat and beautiful. You're beautiful but you're fat, and you're in an LDR. Come to think of it, who will he choose? Is it you who's living away or the person beside him who likes him? If you're sexy and beautiful, your boyfriend won't be attracted to others. He will even be more excited to meet you in person, in the Philippines or elsewhere.

Lastly, Cathy's final advice is on being responsible. This means how one should be a caring and thrifty person. She then contrasts these values with the behaviours of other Filipinas:

> Foreigners often search for Filipina because those foreigners who marry a Filipina are happy because they're responsible … It's frustrating to see some Filipino women who are being complained by their partners because they're a gold digger, not taking care of their husband and children, and splurges money. That's why sometimes we're roped into a bad judgment.

In these statements, she shows how Filipino women in an interracial relationship can assert their agency through their behaviours. However, it is interesting to note that the advice also reinforces a certain type of stereotype. First, she particularly highlights the need for self- improvement – being sexy and beautiful – to nurture a healthy and long-lasting relationship. Further, she also highlights the domesticated figure of a Filipina, noting a caring wife and mother. Across these representations, we see how agency becomes relational, which practices often feed a patriarchal system.

Intimate engagements with platform-specific tactics

We observed that YouTubers deployed a range of strategies that shape the branding and promotion of their videos. We particularly identified these strategies by looking at the "About" section of the YouTube channel, prompt for subscription, the call-to-action statement to facilitate continued online engagement, as well as promotion of multiple platforms used to promote content. These techniques contribute to the YouTubers' relationship-building tactics with their online audiences.

The "About" section of the YouTube channel is crucial in the branding of the content. Among the videos that we analysed, four YouTube channels had descriptions about their relationship. For example, Triple Hearts showed some information about Bonnie and Patrick:

> MABUHAY 🖤 Hello! Welcome to our life. Patrick is from overseas while Bonnie is from the Philippines. We met in the Philippines in 2014. Since then, we have been in long distance relationship for 5 years until we decided to close the distance in April 2019. We got engaged on 24th of November 2019 and getting married soon! HOW EXCITING!!!! :)

Similarly, Team Power Hugs highlights their relationship status in the about section of the YouTube channel, "Hey guys! Welcome to our channel! Here you can follow our (LDR) life. We are still in a Long-Distance Relationship, but we hope to close the distance in 2019!" These statements essentially brand the contents of the YouTube channels as representations of an interracial and long-distance relationship. More importantly, the channels were also framed as a source for advice and tips on navigating such relationships. This is pointed out in the about section of Cathy Lim, "I just would like to share our LDR story and to inspire some couples who have the same situation."

The framing of the contents was operationalised in an entrepreneurial context. Among the ten YouTube contents that we examined, three YouTubers were explicit in presenting their YouTube channel as open to monetisation. They articulated this in two ways. First, they presented their brand through their interests and activities, moving away from the narrative of being in an interracial relationship. This was seen in Jasmine's framing of her channel, which has grown to amass over one million subscribers: "Hi Everyone, Welcome to my channel. I am Jasmine also known as Loving Jas. I am a vlogger/digital influencer with a passion for beauty, fashion and everything lifestyle." This was similar to Cathy Lim's channel description, "Now, I'm already a mom, a mom vlogger, influencer with a passion for beauty, fashion, lifestyle and love sharing all of my mom tips and adventures with you guys. From cooking to a day in the life vlogs, I upload videos of tutorial, tips and daily basis of my day-to-day life along with promotional videos." Further, both channels of Jasmine and Cathy accept business inquiries. As noted in Jasmine's page, "For collaborations and business inquiries: (email) ... Want to send something? (address)." In Cathy's channel, as presented, "For business inquiries: (email)." These strategies primarily show that the YouTube channels are monetised by the YouTubers who curate and visibilise their interracial relationship or life overseas in general.

YouTubers crafted ways of sustaining their audiences and subscribers. They did this in two ways. First, they used call-to-action statements to encourage their viewers to subscribe to their channel. For instance, Regine of *Happy Place* ended her video with the following statement, "Thank you guys for watching. Please like and share if you haven't already. Don't forget to click the subscribe and bell button so you'll be updated for our uploads. Similarly, Jasmine closed her storytime video with such call-to-action, "... please do not forget the subscribe button below. You'll be notified of my videos daily if you hit that subscribe button below there so go ahead and please do it. And I hope to see you guys tomorrow. Bye!" Meanwhile, a YouTuber like Cathy Lim highlighted some of the key contents of her channel, "And for those who are looking for more LDR tips, don't forget to check the subscription box." Second, sustaining audience engagement was managed using multiple platforms. For example, several YouTubers embed icons of their social media channels, which their existing and potential subscribers can follow. Nevertheless, these strategies allowed the YouTubers to generate viewers and subscribers in the channel, which can then be translated as metrics for monetisation of the channel and the contents.

Lastly, valuing the viewers was shown through statements of appreciation among YouTubers, which is considered as a form of communicative intimacy (Abidin, 2015). For instance, at the end of her video, Jasmine states, "I just want to say thank you to all those who are supporting us, you guys are the best. Thank you for always watching my videos. Thank you for always watching our vlogs. Guys thank you. I always say this but I mean it. You guys are the best." Regine of *Happy Place* had the same statement, "And again guys, thank you for watching and we'll see you in our next video. Anyeong!" Notably, YouTubers showed how they value their viewers by encouraging them to suggest topics for the contents. For example, Cathy Lim said, "For those who want to make suggestions, please use the suggestions box. That's for you. Or you can follow me in IG (Instagram), or search my name on Facebook. You can also send a private message if you don't want to be exposed in the comment section." Regine of *Happy Place* had a similar but shorter statement, "Also comment down below if you have any questions for us." Meanwhile, in Jasmine's video, giving a thumbs up for the video was encouraged, "Anyway thank you so much for watching this video. I hope you enjoy it. Please do give this video a thumbs up if you enjoyed watching. And let me know if you want to see more storytime from this channel."

Contradictions of Agency in the Brokerage of Interracial Intimacies

The ten videos that we analysed showed the experiences of Filipino women and their foreign partners in a long-distance relationship. These videos mostly presented the perspective of Filipino women. We argue that these women acted as brokers of meaningful and successful interracial relationships, which have been articulated through a range of representations, practices and platform-specific strategies. In the first instance, the narratives begin with detailing how the couple met via matchmaking and online sites. The interactions progress and further develop using other communications platforms. A common key highlight is the first physical meetup of the couple, which shows how co-located interactions are performed and embodied through playful and intimate activities, including eating together, cuddling, and visiting several tourist spots. These moments are succeeded by the couple's return to a long-distance arrangement again by utilising digital communication technologies. Eventually, the video concludes with the happy reunification of the couple in the foreign national's home country. All these constitute the brokering of information crucial for constructing an aspiration of an interracial relationship, offer evidence of its possibility, solidifying its value for an ordinary Filipina as an "image of an ideal relationship," and sustaining the connection of multiple actors that desire and exchange their personal stories of intimacy through comment loops.

The narratives are like fairy tales (Rondina, 2004) yet offer a different style and trajectory. Scholars have argued that promotional materials of online or matchmaking sites often show a happy ending of online meet ups (Constable, 2003; Rondina, 2004). However, we noticed that prior to the happy ending of the story, individuals presented the "effort" needed to overcome the challenges, and ultimately, the rewards of forging and maintaining a long distance and interracial relationship. All these connect powerfully to the aspirations held by many of their avid subscribers – to find a similar relationship in the future. The Filipina YouTubers also broker women's positionality, challenges, and agentic practices in the relationship. As presented, while Filipino women visibilised the intimate exchanges on a messaging application, Skype, and other online channels, they also expressed the sadness and frustration of living apart, as shown in the use of images, of saying goodbye in the airport or by incorporating sad face emojis. Significantly, these fairy tale narratives attempt to challenge the notion of commodified bodies of Filipino women in matchmaking sites. In the videos, relationships and interactions are framed as authentic, which is supported by a range of evidence, including screenshots of exchanges, frequency of contact, and the interaction of the foreign national with the Filipina's friends and family. As such, the story provides an important vantage point on the ways in which Filipino women present their agency in an online space. Here, online representations are countererotics (Constable 2012) or challenge the stereotypical notions of Filipino women who met their foreign partners online (Constable 2003).

We approach the showcase of a fruitful interracial relationship as form of brokerage. In the first instance, the videos show the pathway to overcoming a long-distance relationship and enabling physical togetherness. Furthermore, the Filipina YouTubers construct and curate their persona as someone who is caring, enduring, and domesticated, appealing to cultural expectations in Philippine society. They also framed the character of their partner, someone who is kind, caring and genuine. Nonetheless, the YouTubers also create aspirations for their viewers by showcasing their most personal, intimate and everyday activities, which Abidin (2017) identifies as calibrated amateurism. However, several of these videos are operationalised in an entrepreneurial frame.

Several of the videos incorporated branding and monetisation strategies, indicating the labour additionally embedded in making their representations of the story believable and relatable. The YouTubers not only present strategies or "how to" in dealing with a long distance and interracial relationship. As a form of brokerage, they also sell, implicitly or explicitly, the potentiality of a good life through interracial relationships. This capacity, we argue, is influenced through the Philippines' colonial past. The narratives presented the process of cultural whitening, through the institutional frame of marriage (Arnado, 2019), which also conveys imaginaries of stability, mobility, and security. Nevertheless, through the YouTubers' curation of their intimate and interracial relationships, thumbs up, views, and subscriptions are generated, and these outcomes are constitutive of profitability in an online space of YouTube. Indeed, echoing the contentions of Constable (2009), online practices reflect the commodification of lived and intimate experiences.

Brokerage in online spaces is wrapped in contradictions. Certainly, the success stories of the Filipina YouTubers counter the stigma or sexualisation of women. They broadcast their lives with a sense of agency, as reflected in their story making, visualisation, and navigation of online spaces. However, in a marketised environment, they generate profit from their online practices, which essentially engage audiences and complement the platform's operations. This speaks to being an entrepreneurial and self-driven woman in digital cultures (Banet-Weiser,

2012). Significantly, representations are political. The self-portrayal of women tends to reinforce a caring and altruistic individual, appealing to the demands and expectations of the viewers who contain women in certain categories. In some cases, the viewers celebrate the stereotypical role of Filipino women in an interracial relationship – someone who is there to care for the man and deliver domestic work (Cabalquinto & Soriano, 2020). Here, Filipino women can be confined to domestic subjectivities, reinforcing the historically situated gendered, racialised, and classed structures in Philippine postcolonial reality. Significantly, such representation becomes a modality for profit-making because this is what is demanded by the viewers whose views about individuals are tied to hierarchy (Sobande, 2019). Nevertheless, we argue that the brokering interracial relationships signal the negotiations of Filipino women in navigating the constraints and hierarchies in a postcolonial, neoliberal, and digital Philippine society.

The Possibilities and Economies of Interracial Intimacies

This chapter has shown the dynamics of brokering interracial relationships among Filipino women and foreign nationals on YouTube. By analysing ten YouTube videos of Filipino women in an interracial relationship, we highlight how intimate, personalised, and creative narratives serve as a negotiation of women's positionality in postcolonial and neoliberal contexts. To begin with, using the concept of countererotics, we display how the imagery of a Filipina counters the sexualised and commodified figure. Further, men in the videos are represented as being kind, caring, and accepting. Notably, the narratives are presented in a fairy tale style, yet highlighting the possibility of a successful relationship as initiated and sustained by using a range of online channels and authentic exchanges. However, using a "cultural whitening" frame, we argue that showcasing intimate and interracial re-lationships on YouTube has the tendency to perpetuate the imaginaries of certain racialised, gendered and classed bodies. In some of the narratives of the YouTubers, women's subjectivity in a relationship is constructed and policed to cater to the male and colonial gaze. One has to cater to the desires of men in order to nurture a healthy relationship. Further, women's stories serve as launch pads for stirring imaginaries of happiness, security and social mobility. These online and affective representations articulate romanticised versions of intimate relationships, attract viewership, stir online engagement and generate profit-making. In a digital environment, curated and popular contents are amplified by the platforms' algorithms and entrepreneurial system. Nevertheless, the brokerage of interracial ties on YouTube underscore the tensions and negotiations in digital cultures, showing how agency, visibility, and branding strategies are co-opted and negotiated in networked, marketised and postcolonial spaces.

References

Abidin, C. (2015). Communicative ♥ intimacies: Influencers and perceived interconnectedness. *Ada: A Journal of Gender, New Media, and Technology, 8,* 1–16. doi:10.7264/N3MW2FFG

Abidin, C. (2017). #familygoals: Family influencers, calibrated amateurism, and justifying young digital labour. *Social Media + Society, 3*(2), 1–15. doi:10.1177/2056305117707191

Abidin, C. (2018). *Internet celebrity: Understanding fame online* (First ed.). Emerald Publishing Limited.

Aguilar, F. J. (2014). *Migration revolution: Philippine nationhood and class relations in a globalized age.* Ateneo de Manila University Press.

Amy, L. E. (1997). The mail-order bride industry and immigration: Combating immigration fraud. *Indiana Journal of Global Legal Studies, 5*(1), 367–374.

Aquino, K. (2018). *Racism and resistance among the Filipino diaspora: Everyday anti-racism in Australia.* Routledge.

Arnado, J. M. (2019). Cultural whitening, mobility and differentiation: Lived experiences of Filipina wives to white men. *Journal of Ethnic and Migration Studies,* 1–17. doi:10.1080/1369 183X.2019.1696668

Association of Internet Researchers (AoIR) (2012). Ethical decision-making and Internet research: Recommendations from the AoIR ethics working committee (version 2.0). Retrieved 12 January 2020, https://aoir.org/reports/ethics2.pdf

Association of Internet Researchers (AoIR) (2019). Internet research: Ethical guidelines 3.0. Retrieved 10 January 2020, https://aoir.org/reports/ethics3.pdf

Banet-Weiser, S. (2011). Branding the post-feminist self: Girls' video production on YouTube. In M. C. Kearney (Ed.), *Mediated girlhoods: New explorations of girls' media culture* (pp. 277–294). Routledge.

Banet-Weiser, S. (2012). *Authentic TM: The politics and ambivalence in a brand culture*: New York University Press.

Bauman, Z. (2000). *Liquid modernity.* Polity Press.

Baym, N. (2015). Connect with your audience! The relational labour of connection. *Communication Review, 18*(1), 14–22. doi:10.1080/10714421.2015.996401

Burgess, J. (2007). Vernacular creativity and new media (Doctoral dissertation), Queensland University of Technology, Brisbane, Australia. Retrieved 26 October 2020, http://eprints. qut.edu.au/16378/1/Jean_Burgess_Thesis.pdf

Burgess, J., & Green, J. (2018). *YouTube* (Second ed.). Polity Press.

Cabalquinto, E. C., & Soriano, C. R. (2020). 'Hey, I like ur videos. Super relate!' Locating sisterhood in a postcolonial intimate public on YouTube. *Information, Communication & Society, 23*(6), 892–907. doi:10.1080/1369118X.2020.1751864

Cabañes, J. V. (2014). Multicultural mediations, developing world realities: Indians, Koreans and Manila's entertainment media. *Media, Culture & Society, 36*(5), 628–643.

Cabañes, J. V. (2019). Information and communication technologies and migrant intimacies: The case of Punjabi youth in Manila. *Journal of Ethnic & Migration Studies, 45*(9), 1650–1666. do i:10.1080/1369183X.2018.1453790

Cabañes, J. V., & Collantes, C. (2020). Dating apps as digital flyovers: Mobile media and global intimacies in a postcolonial city. In J. V. Cabañes & L. Uy-Tioco (Eds.), *Mobile media and social intimacies in Asia: Reconfiguring local ties and enacting global relationships* (pp. 97– 114). Springer.

Choy, C. C. (2003). *Empire of care: Nursing and migration in Filipino American history.* Duke University Press.

Commission on Filipino Overseas (2018). Statistical profiles of spouses and other partners of foreign nationals. Retrieved 10 May 2020, from https://cfo.gov.ph/statistics-2/

Constable, N. (2003). *Romance on a global stage: Pen pals, virtual ethnography, and "mail-order" marriages.* University of California Press.

Constable, N. (2005). A tale of two marriages: International matchmaking and gendered mobility. In N. Constable (Ed.), *Cross-border marriages: Gender and mobility in transnational Asia* (pp. 166-186). University of Pennsylvania Press.

Constable, N. (2006). Brides, maids, and prostitutes: Reflections on the study of 'trafficked' women. *Portal: Journal of Multidisciplinary International Studies, 3*(2), 1–25. doi:10.5130/portal.v3i2.164

Constable, N. (2009). The commodification of intimacy: Marriage, sex, and reproductive labour. *Annu. Rev. Anthropol., 38,* 49–64. doi:10.1146/annurev.anthro.37.081407.08513

Constable, N. (2012). Correspondence marriages, imagined virtual communities, and countererot-ics on the Internet. In P. Mankekar & L. Schein (Eds.), *Media, erotics, and transnational Asia* (pp. 111–138). Duke University Press.

Del Vecchio, C. (2007). Match-made in cyberspace: How best to regulate the international mail order bride industry *Columbia Journal of Transnational Law, 46*(1), 177–216.

Dobson, A. S., Robards, B., & Carah, N. (2018). Digital intimate publics and social media: Towards theorising public lives on private platforms. In A. S. Dobson, B. Robards, & N. Carah (Eds.), *Digital Intimate Publics and Social Media* (pp. 3– 27). Palgrave Macmillan.

Ellison, N., Heino, R., & Gibbs, J. (2006). Managing impressions online: Self-presentation processes in the online dating environment. *Journal of Computer-Mediated Communication* (2), 415.

Garcia, C. R. (2011). Exploring the realities of Korean-Filipino marriages. Retrieved 12 June 2021, htt-ps://news.abs-cbn.com/global-filipino/11/07/11/exploring-realities-korean-filipino-marriages

García-Rapp, F. (2017). Popularity markers on YouTube's attention economy: The case of Bubz-beauty. *Celebrity Studies, 8*(2), 228–245. doi:10.1080/19392397.2016.1242430

Giddens, A. (1991). *Modernity and self-identity: Self and society in the late modern age.* Polity Press in association with Blackwell Publishing Ltd.

Glatt, Z., & Banet-Weiser, S. (2021). Productive ambivalence, economies of visibility and the political potential of feminist YouTubers. In S. Cunningham & D. Craig (Eds.), *Creator culture: Studying the social media entertainment industry* (pp. 39–56). New York University Press.

Glenn, E. N. (2009). Consuming lightness: Segmented markets and global capital in the skin-whitening trade. In G. E. N. (Ed.), *Shades of difference: Why skin color matters* (pp. 166–187). Stanford University Press.

Gonzalez, V. V., & Rodriguez, R. M. (2003). Filipina.com: Wives, workers and whores on the cyberfrontier. In R. Lee & S.-l. C. Wong (Eds.), *Asian America.Net: Ethnicity, Nationalism, and Cyberspace* (pp. 215–234). Routledge.

Guevarra, A. R. (2010). *Marketing dreams, manufacturing heroes: The transnational labour brokering of Filipino workers.* Rutgers University Press.

Hjorth, L. (2011). It's complicated: A case study of personalisation in an age of social and mobile media. *Communication, Politics & Culture, 44*(1), 45–59.

Hobbs, M., Owen, S., & Gerber, L. (2017). Liquid love?: Dating apps, sex, relationships and the digital transformation of intimacy. *Journal of Sociology, 53*(2), 271–284. doi:10.3316/informit.407452887319646

Hunter, M. L. (2005). *Race, gender, and the politics of skin tone.* Routledge.

Kusel, V. I. (2014). Gender disparity, domestic abuse and the mail-order bride industry. *Albany Government Law Review, 7*(1), 166–186.

Lange, P. (2007). Publicly private and privately public: Social networking on YouTube. *Journal of Computer-Mediated Communication, 13*(1), 361–380.

Lee, D. R. (1998). Mail fantasy: Global sexual exploitation in the mail-order bride industry and proposed legal solutions. *Asian Law Journal, 5*(1), 139–179.

Licoppe, C. (2014). *Location awareness and the social-interactional dynamics of mobile sexual encounters between strangers: The uses of Grindr in the gay male community.* Paper presented at the Social Lives of Locative Media, Swinburne University of Technology.

Lindee, K. M. (2007). Love, honor, or control: Domestic violence, trafficking, and the questions of how to regulate the mail-order bride industry. *Columbia Journal of Gender and Law, 16*(2), 551–612.

Lobato, R. (2016). The cultural logic of digital intermediaries: YouTube multichannel networks. *Convergence, 22*(4), 348–360. doi:10.1177/1354856516641628

Madianou, M., & Miller, D. (2012). *Migration and new media: Transnational families and polymedia.* Routledge.

Marwick, A. E. (2013). *Status update: Celebrity, publicity, and branding in the social media age.* Yale University Press.

Parreñas, R. S. (2011). *Illicit flirtations: Labour, migration and sex trafficking in Tokyo.* Stanford University Press.

Quiroz, P. A. (2013). From finding the perfect love online to satellite dating and 'loving-the-one you're near': A look at Grindr, Skout, Plenty of Fish, Meet Moi, Zoosk and Assisted Serendipity. *Humanity & Society, 37*(2), 181–185. doi:10.1177/0160597613481727

Rafael, V. (2000). *White love and other events in Filipino history.* Duke University Press.

Raun, T. (2018). Capitalizing intimacy: New subcultural forms of micro-celebrity strategies and affective labour on YouTube. *Convergence, 24*(1), 99–113.

Robinson, K. (1996). Of mail-order brides and 'boys' own' tales: Representations of Asian-Australian marriages. *Feminist Review*(52), 53–68. doi:10.1057/fr.1996.7

Roces, M. (2009). Prostitution, women's movements and the victim narrative in the Philippines. *Women's Studies International Forum, 32*, 270–280. doi:10.1016/j.wsif.2009.05.012

Rodriguez, R. M. (2010). *Migrants for export: How the Philippine state brokers to the world.* The University of Minnesota Press.

Rondina, J. (2004). The e-mail order bride as postcolonial other: Romancing the Filipina in web based narratives. *Plaridel: A Philippine Journal of Communication, Media, and Society, 1*(1), 47–56.

Rondilla, J. (2009). Filipinos and the color complex, ideal Asian beauty. In G. E. N. (Ed.), *Shades of Difference: Why skin color matters* (pp. 63–80). Stanford University Press.

Rondilla, J. L., & Spickard, P. (2007). *Is lighter better?: Skin-tone discrimination among Asian Americans.* Rowman & Littlefield Publishers.

Saroca, C. (2007). Filipino women, migration and violence in Australia: Lived reality and media image. *Kasarinlan: Philippine Journal of Third World Studies, 21*(1), 75–110.

Sobande, F. (2019). Constructing and critiquing interracial couples on YouTube. In G. D. Johnson, K. D. Thomas, A. K. Harrison, & S. A. Grier (Eds.), *Race in the Marketplace*. (pp. 73–85). Palgrave Macmillan.

Spivak, G. C. (1999). *A critique of postcolonial reason: Toward a history of the vanishing present*. Harvard University Press.

Starr, E., & Adams, M. (2016). The domestic exotic: Mail-order brides and the paradox of globalized intimacies. *Signs: Journal of Women in Culture & Society, 41*(4), 953–975. doi:10.1086/685480

Strangelove, M. (2010). *Watching YouTube: Extraordinary videos by ordinary people*. University of Toronto Press, Scholarly Publishing Division.

Tadiar, N. X. M. (2004). *Fantasy production: Sexual economies and other Philippine consequences for the new world order*. Hong Kong University Press.

Tolentino, R. B. (1996). Bodies, letters, catalogs: Filipinas in transnational space. *Social Text*(48), 49–76. doi:10.2307/466786.

van Dijck, J. (2013). *The culture of connectivity: A critical history of social media*. Oxford University Press.

Zimmer, M. (2010). 'But the data is already public': On the ethics of research in Facebook. *Ethics and Information Technology*, 12, 313–325.

Zug, M. A. (2016). *Buying a bride: An engaging history of mail-order matches*. New York University Press.

5 Labour

Abstract

This chapter offers a critical insight on digital and flexible labour and the role of YouTube in brokering labour relations and economic aspirations in the digital economy. It focuses its attention on the "skill-making" content and practices of Filipino platform labour influencers on YouTube, where they showcase their capabilities to obtain a captive market and attain celebrity status as "global knowledge workers." Using YouTube provides an opportunity for platform labour influencers to deliver training and support to aspiring platform workers who seek to earn dollars while working at home, while also crafting imaginaries and ideals of success in the platform economy. The chapter shows that YouTubers, through the brokerage of skills and promotion of the viability of platform labour, perform the role of matchmaker between aspiring workers and digital labour platforms while simultaneously advancing the broader visions of digital capitalism.

Keywords: platform labour, digital labour brokerage, skill-making, aspirational labour, gig economy, cloudwork

On a YouTube video, Carlo beams with pride as he announces to his thousands of subscribers a new milestone: he has made his second million as an online freelancer. Displaying his monthly dollar earnings, a new car, as well as a modern workstation, Carlo is ecstatic to proclaim that he has made the best decision to leave his full-time call centre job for platform work, challenging his subscribers that they could achieve the same success with the right recipe of hard work, persistence, and entrepreneurialism. He has worked hard to achieve his current state, starting as a book chapter writer for a small client clinched from a labour platform way back in 2011 and now a digital marketing "guru," working with a number of clients from the popular labour platform *Upwork*, and recognised as the "coach" for hundreds of Filipino platform workers and aspirants. In 2016, Carlo started

Soriano, Cheryll Ruth and Earvin Charles Cabalquinto: *Philippine Digital Cultures: Brokerage Dynamics on YouTube*. Amsterdam: Amsterdam University Press, 2022
DOI: 10.5117/9789463722445_CH05

to organise training sessions for friends who wanted to learn how to start with online freelancing after noticing and being inspired by a substantial growth in his earnings. Realising that training is a possible mechanism to help aspiring Filipino platform workers and at the same time a potential source of additional income, he began to organise these sessions regularly, initially with a few aspirants. Now Carlo speaks regularly at freelancer events, organises paid coaching classes, hosts his YouTube channel on online freelancing, and runs a Facebook community composed of his former and current trainees and their network of friends. His YouTube channel is dedicated to regular videos on how to survive and thrive in the freelancing world: from teaching about specific skills, to how to set up an online office, how to manage a foreign client, and how to invest one's earnings.

Burdened by employment woes such as infrastructural immobility and low wages, countless Filipino professionals like Carlo are found to be migrating to online platform labour in exchange for autonomy, spatial flexibility, and the possibility for higher earnings. Meanwhile, those not meeting the eligibility requirements of traditional labour – whether due to physical disabilities or educational attainment – may now also justly compete for jobs by curating an attractive portfolio that highlights their skills. As briefly discussed in Chapter 1, platform labour is likewise being championed by the Philippine government as a solution to bridge its rural employment gap. Platform workers (more commonly called *online freelancers* in the Philippines), mostly located in the Global South, can now directly obtain "gigs" through online labour platforms and microwork intermediaries such as Upwork, Onlinejobs.ph, Rev, Fiverr, or Freelancing.com. The Philippines has become one of the largest labour supplying countries in these platforms (Graham, Hjorth, & Lehdonvirta, 2017, p. 142), now ranked first among the fastest growing freelancing markets in the world with 208% growth in 2020 (Payoneer, 2020). In contrast to business process outsourcing (BPO) such as call centre work, digital labour platforms allow business processes to be outsourced without the mediation of formal BPO companies (Graham et al., 2017). Platforms perform the labour matching role where clients or buyers of work (mostly located in high-income countries) post jobs in these platforms and aspiring workers (mostly located in low-income countries) can bid on them directly (McKenzie, 2020; Wood et al., 2016).

This chapter focuses on the rise of platform labour brokers (Soriano, 2020) on YouTube (and the informal economy underlying them) who attract and train Filipino platform workers into this labour market, therefore performing a crucial, but under-noticed brokerage role between workers and platforms, and among workers. The chapter will show that for Filipinos, YouTube has become a crucial site for the circulation of labour aspirations in the digital economy, also underscored by the Philippine postcolonial condition. These digital labour brokers on YouTube are playing an important role in facilitating the local popularity and viability of platform labour and at the same time pushing norms among local workers in this industry.

The need for influencers, locally called coaches, trainers, or peer mentors, embodies the transaction of ambiguity underlying digital labour in the Philippines. In the advent of abundant information infrastructures and "flexible" work environments, it is assumed that intermediaries will be bypassed in electronic markets as workers can do away with traditional hiring and employment procedures as well as gatekeepers and connect to potential clients directly. Yet, we see an emerging category of digital labour intermediaries on social media who are playing a significant role in the expansion and continued uptake of digital platform labour in the country. In this chapter, we examine the transactional nature underlying this producer-audience relationship, the activation of trust and influence through personalised practices and mediated encounters, as well as the power dynamics underlying the digitally mediated symbolic and material power taking place between them and their respective teams and subscribers.

Platform Labour in the Philippines

Platform labour is celebrated by the Philippine government as a viable solution to unemployment and emerges as a highly attractive work option especially as employment conditions in the country are fraught with socio-economic tensions. The popularity of platform labour is expanding alongside the COVID-19 pandemic which has resulted in thousands of labour displacements both in the local economy and among Filipino migrant workers who were forced to return home. Marketed as a flexible and competitive source of income, platform labour has also been attracting Filipino workers who have trouble in coping with the conditions surrounding older employment models such as BPO, overseas labour migration, or in supplementing casual and unstable employment elsewhere. Government's over-optimism toward digital labour can be seen in parallel to Richard Florida's notion of the rise of the "creative class," a group of socio-economic subjects who had jobs based on creativity and individual talent and who could usher cities into a new era of economic development and prosperity while addressing the unemployment gap (Florida, 2014). Many Filipino workers find platform work appealing as it provides them with new work opportunities to earn dollars, facilitates entry into the "global tech workplace," and offers a flexible work arrangement that allows them to overcome the challenges of commuting while giving them an opportunity to be economically productive while spending valuable time with their families at home (Soriano & Cabañes, 2020a, 2020b). A characteristic that attracts many Filipinos to platform labour is their English proficiency, developed through an American education system and postcolonial culture that privileges English proficiency as a ticket to success (Soriano & Cabañes, 2020a). Our interviews with platform workers also show that digital labour has created aspirations for workers in the countryside, as well as those with disabilities, who are often ineligible for local employment, to join the global knowledge workforce. Yet while workers are able to gain substantial benefits from platform labour, research on digital worker experiences highlight its problematic realities, which include increasing levels of anxiety over labour insecurity, financial and career instability, limited bargaining capacity over clients and platforms, physical stress, and isolation – all of which challenge the over-optimism accorded to digital labour by government and platform promotions (Hesmondhalgh & Baker, 2010; Graham, Hjorth et al., 2017; Lehdonvirta, 2016; Soriano & Cabañes, 2020b).

Despite strong government promotion of digital labour as a solution to unemployment, mechanisms for supporting workers are limited. For BPO-related jobs such as call centre work, foreign companies operating in the country have institutionalised recruitment and employment mechanisms (Padios, 2018), with human resource officers guiding aspiring workers in the hiring and employment processes and with employee handbooks outlining details for promotion possibilities and navigating day to day operations. For labour migration, several private and public institutions such as migration agents (Lindquist, Xiang & Yeoh, 2012; Shresta & Yeoh, 2018) and government departments have been set up to assist workers aspiring to migrate overseas in terms of employment seeking, making sense of local and foreign bureaucracies, expectation-setting, salary identification, taxation, or welfare protection (Rodriguez, 2010). By contrast, many new entrants and aspiring digital freelance workers try to learn the ropes through YouTube or by directly taking on gigs from different platforms, which can involve direct experiences of scams, client abuse, acceptance of low rates, or frustration over the inability to obtain well-earning gigs from labour platforms.

Given the tension between the promising opportunities posed by platform labour and the lack of institutional mechanisms to help workers navigate the ambiguous digital platform environment, these influencers perform a brokerage role in various social media platforms and freelancer circles as they usher workers toward aspirations of stable six-digit incomes, crafting an attracting portfolio, gaining loyal foreign clients, and obtaining a sense of self-fulfilment by becoming economically productive while performing their nurturing roles at home. Most of them current or former digital platform workers who achieved some level of success, they know the realities and ambiguities of shifting into platform labour from regular full-time employment, the pernicious conditions yielded by the platforms' design, and have been exposed to the difficulties of working with foreign clients (Soriano & Panaligan, 2019). They rise as an "elite" group of workers because they capitalise on their experiences and exposure to the challenges of digital labour and translate these into aspirational narratives which they also monetise.

Labour Brokerage

Online freelancing influencers on YouTube represent a contemporary manifestation of labour brokerage in the Philippines, although the intermediary role of labour brokers has been persistent historically. Amid labour-only contracting which dates back to the Spanish colonisation from 1565–1898, a labour arrangement called the *cabo* system was used in negotiating and supplying labour for various kinds of industries and especially in agricultural farmlands or *haciendas* that hired workers on a seasonal basis (IBON Foundation, 2017; Kapunan, 1991; Silarde, 2020). Because Filipinos were deprived of the right to own land and had no choice but to work with agricultural contractors, *cabos* performed the role of middlemen and acted as negotiators between workers and contractors, determining how much workers would be paid for a specific scope of work or in helping workers determine where work was available (Kapunan, 1991, p. 326; Soriano, 2021). *Cabos* deducted a percentage fee for their brokerage role (Kapunan, 1991, p. 328) and this set-up thrived under patronage structures and the "paternalistic orientation of Filipinos" as they were also seen as the "fathers of a big family," providing workers information and advice beyond work-related matters and with the labour transactions fusing with social and community relations (Kapunan, 1991, p. 327). After over three centuries of Spanish rule, the American colonial system suppressed the *cabo* system and instead promoted the adoption of trade unionism. However, historical records show that the *cabos* merely took on leadership roles in trade unions to sustain their influence, capitalising on networks that they had previously established (Silarde, 2020).

Beyond the Spanish and colonial rule, labour brokerage entities have taken over the place of *cabos* by directly or indirectly intermediating for workers in their search for employment. With the Philippine government pushing for labour export policy and celebrating labour migrants as modern-day heroes, many Filipinos aspired to work overseas and sought guidance on how to navigate national and global migration processes and requirements (Rodriguez, 2010; Soriano, 2021). Although national government agencies have been set up to support overseas labour migration, and with a growing number of Filipinos aspiring to clinch a job overseas to earn dollars, "migration agents" (Lindquist, Xiang & Yeoh, 2012; Shresta & Yeoh, 2018) or "private recruitment agencies" (Lin, Lindquist, Xiang et al., 2017) persisted because they offered key services that these government agencies do not sufficiently provide. These labour migration brokers do not just put together ambiguous labour migration information for aspirants, some of them help directly facilitate a match between Filipino workers and foreign contractors, of course deducting significant fees in return for the transaction.

Like platform labour, the labour migration process involves significant ambiguities. One needs to figure out and make sense of the numerous documentary requirements and obtain the necessary approvals and signatures and navigate both national and foreign bureaucracies. As we have discussed in Chapter 2, for an ordinary aspiring worker without access to informational resources, the service offered by brokers is indispensable for helping aspiring workers understand labour migration procedures, simplify the complicated steps and requirements, make sense of the different agencies and their documentary requirements, and mediate in complex transactions. Secondly, like platform labour, migrant labour is laden with scams and abusive employers. Workers know about these scam stories from documented evidence about Filipinos ending up working with clients who refuse to pay or who physically or violently abuse workers. Some of these agents promise workers to match them with well-screened clients who are respectful of labour and human rights, although this does not always translate in reality. According to Shresta and Yeoh (2018), it is the brokers' wealth of information on local and global labour migration processes that legitimise their work, allowing them to charge significant amounts of money from Filipino migrant workers with the promise that after years of overseas work, they would eventually be able to recuperate these fees. And yet, it is also well documented how many Filipino migrant workers carry the debts they incur from migration brokers for many years, sometimes compelling them to remain as migrant workers for extended periods of time.

At the core of brokerage – from the *cabo* to the migration agent systems and to YouTube platform influencers – is informality. We pertain
to "informality" in terms of "economic transactions outside the view and
formal control of governments and corporations" (Kelly, 2001) to highlight
and contrast this with transactions enacted within formal regulatory
structures. Brokerage can appear as a set of localised, often alternative
services and provisions associated with informal economic transactions
within local infrastructures that facilitate and regulate labour (Shresta &
Yeoh, 2018, p. 671). Brokerage, whether formally (through registered agencies)
or informally (through personal networks), become relevant due to the
government's shortcomings as well as decline in trust in large institutions,
including labour unions, throughout the industrial world (Neilson & Rossiter,
2008) and in the Philippines (Ofreneo, 2013; Soriano, 2021).

YouTube Influencers: Brokering Platform Labour Ambiguity

This chapter draws from over four years of interviews with forty-one
prominent digital platform labour influencers and Filipino freelance workers
as well as an analysis of YouTube videos of digital labour influencers on
YouTube. Face to face and online interviews with workers and influencers
were conducted from January 2017 to December 2019, and participants were
recruited through meetups in freelancer events. Observations during these
events as well as initial interviews were important for constructing background information about the online freelancing scene in the country and
provided clues on the role of social media influencers in the digital labour
economy. These interviews made it clear to us that emerging influencers
are acting as brokers, playing a crucial role in advancing platform labour
in the country, and that these are conducted through social media and
through meetups and events. We noted the names of prominent coaches
identified by the workers as playing key roles in promoting online freelancing, some of whom were also interviewed for the study. Over the course of
the research, we noticed the emerging importance of online freelancing
channels on YouTube where video content for navigating platform labour
is actively created and shared outside the platform, and specifically into
Facebook, Messenger, and Viber freelancing groups. This prodded us to
analyse the top eight Filipino digital labour influencers on YouTube whose
channels have a range of 22,600 to over half a million subscribers. Although
some of them have relatively fewer subscribers than others, the videos they
produced gained 32,000 to almost 1.5 million views each, indicating that

Table 5.1   Digital labour brokers and sample of videos analysed

Digital labor influencer	Date of joining YouTube	Subscribers	Total no. of videos in channel	Sample of video analyzed	Date published	Video Views
Austin Gabriel Diaz	05/18/2020	59,600	28	*PAANO KUMITA ONLINE (How to earn online?)*	FREELANCING JOBS FOR BEGINNERS	June 27, 2020	64,214
James Tristan Ruiz	04/06/2017	588,000	188	How to Earn 10$ in Online Job in 10–20mins Online Jobs at Home Philippines for Beginners	July 19, 2017 Nov 27, 2019	1,498,554 381,000
Edille Rosario	18/04/2019	84,100	90	*Nag Apply Ako Sa Upwork As A Beginner After 1 Hour May First Client Na Ako (I applied in Upwork as a Beginner and after 1 hour I have a client)*	May 18, 2020	90,220
Demi Bernice	09/06/2015	22,600	127	How To Transition from 9-5 Regular Day Job to Freelancing	June 07, 2019	3,381 views
Mimi Luarca	14/11/2011	356,000	142	Be a Part-time Transcriptionist in REV and Earn 11,000 pesos! Work from Home English Subtitles	October 21, 2019	483,000
Jhazel de Vera	30/06/2015	114,000	234	EARN UPTO $500 / MONTH BY VIEWING IMAGES	Hashingadspace.com Review	Aug 13, 2019	464,000
Eraldo Vlogs	09/01/2013	131,000	75	PAANO KUMITA SA FACEBOOK? (How to earn on Facebook) Ad Breaks (In-Stream Ads) Tutorial	July 23, 2019	408,783
Sheena Santos	04/10/2011	50,600	119	HOW I EARNED $5,335 ON FIVERR?	SECRETS REVEALED	Online Part Time Jobs How To Make Money On Fiverr Without Skills – How To Make Money On Fiverr For Beginners (2020)	May 16, 2020 Aug 12 2019	32,977 236,171

their influence as online freelancing influencers is substantial. We then conducted a textual analysis of the videos and channels of these Filipino digital labour influencers (Refer to Table 5.1 for the brokers and sample YouTube videos analysed) to examine their narrative and engagement styles.

Platform labour and micro-influencer work align – their work in online platforms involve enacting an "entrepreneurial spirit," embodying and at the same time promoting resilience, passion, and hard work needed to survive in a neoliberal economy (Marwick, 2013). Digital labour influencers initially attract a following through face-to-face coaching sessions beginning with a small pool of aspiring freelancers. As they start to earn a name, these brokers would be invited to speak during online freelancing events or start YouTube channels and Facebook groups which allow them to develop a network of trainees. They then become the main connective node for these members, creating an associational network among their subscribers, including those who have been coached by the same broker. It is when they amass a group of subscribers and mentees that they are able to fully enact their brokerage role. YouTube content creators covered by the study appear to be middle aged middle-aged males and females (at least in their self-presentation or persona).

Platform labour influencers engaged diverse strategies of self-branding, skill-making, community building, and profit generation to position themselves in the digital labour market and generate followers and clients, while also crafting strategic networking activities in a hybrid online or offline space (Abidin, 2015; Soriano & Panaligan, 2019). Most of the content and videos produced by them followed the self-tutorial format and other "self-branding" strategies such as sharing personal experience, highlighting one's "authenticity," as well as using video branding styles in the opening and closing credits of their videos (Abidin, 2015). They also shared their personal freelancing journey of success, enveloped between practical tips, tricks, and recommendations. The notion of self-branding implies that micro-celebrities need to differentiate themselves from other influencers by creating a relatable and unique persona or character (Abidin, 2017). Consistent with the work of micro-celebrities, visual and audio styles of self-branding included the use of the colour scheme, font, layout, and music bed, which are consistently used as "branding strategies" across the videos of each influencer (Abidin, 2015; Marwick, 2013; Senft, 2013). The following sections illustrate the brokerage work of these emerging digital labour influencers, which involves: a) addressing digital labour aspirations by illustrating opportunity and highlighting possibility; b) creating a community among Filipino platform workers; and c) cascading and normalising values.

Selling aspirations: Illustrating opportunity and highlighting entrepreneurial strategies

The key role that YouTube brokers play is illustrating the opportunities and possibilities presented by platform labour to aspiring and current Filipino workers. Many aspirants and existing platform workers explore YouTube for resources to help them clinch a platform gig, learn new skills crucial for boosting one's portfolio, or to find strategies to earn better. Some, especially those who have left their full-time physical jobs, search for affirmation that their decision to move to online freelancing can indeed allow them to achieve sustainable income. In turn, brokers' videos play a role in ushering aspiring platform workers towards the vision of what they can be in this digital labour environment and train them with practical strategies to achieve "success."

Being platform workers themselves, these influencers recognise the challenges and ambiguities attached to platform work. These include *labour arbitrage* (or the process of clients hiring the cheapest offer from a large labour supply pool, *labour seasonality* (where demand for work may be high in certain seasons but "dry" in some), navigating the platform's socio-technical system of hiring and rating workers, and the difficulties of managing relationships with foreign clients, among others. Brokers offer a slew of insights and training packages enveloped in their YouTube videos to cascade the vision that despite the challenges, it is possible to achieve the promised successes of platform labour with the "right recipe for success."

For platform workers and especially for new entrants, platform labour is confusing, and the fully mediated form of labour engagement implies that workers need to navigate it without much formal guidance from any institutional authority. These ambiguities include discerning legitimate platforms and clients, building a compelling digital portfolio, figuring out how to bid for a job, how to perform a job online, the relative advantages of different platforms, or how to manage a foreign client and their demands (Soriano & Panaligan, 2019). Strategic pricing is also a common source of ambiguity given the nature of bidding for jobs embedded in the digital labour platform design. This pertains to confusion on how to price one's skills in the job bidding process, which relies on a whole range of considerations such as extent of experience or the nature, size, and duration of the project. For those transitioning from full-time physical employment to full-time platform labour, workers express concerns about the seasonality of work or whether there are mechanisms for negotiation if clients end up unwilling to pay for completed tasks. Importantly, many aspiring workers do not possess college

Figure 5.1   Still from a YouTube video (Santos, 2019). Broker illustrates labour platform possibilities for beginners.

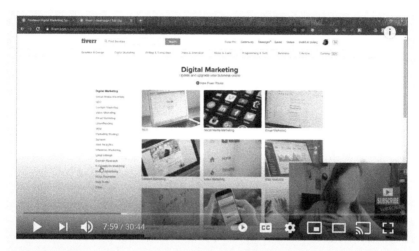

HOW TO MAKE MONEY ON FIVERR WITHOUT SKILLS - HOW TO MAKE MONEY
ON FIVERR FOR BEGINNERS (2020)

degrees nor direct training to perform the tasks offered in platform labour. Clinging on to the promises of platform labour, they try to "learn" these on the go through watching YouTube videos and joining Facebook communities organised by these influencers. In fact, some workers shared with us how they managed to ace projects for which they had no prior experience by watching videos made by "coaches" on YouTube (Figure 5.1)

Brokers acted as authority figures who guided digital workers in navigating their way around digital labour's precarious and ambiguous working conditions. Posts shared on YouTube ranged from "how to begin as an online freelancer," "how to sell one's skills," coaching on specific skills (i.e., website development, social media management, web design, transcription, digital marketing, or virtual assistance, among others), or advice on how to make successful investments of their earnings – which was particularly crucial in addressing concerns tied to labour seasonality attached to online free-lance work. These videos also included how couple freelancers (where both partners do cloudwork) can efficiently work from home while maintaining a fruitful domestic life – and which targeted one of the key aspirations of many Filipino freelancers.

The range of platform labour videos on YouTube presented an interesting contrast to the nature of digital labour training offered by the Philippine government, which focuses more on basic computing and technical skills for aspiring workers in rural communities (E.L. Delfin & M. Fabricante, personal communication, February 5, 2020). For many of the workers we interviewed, successful entry and sustaining platform work required more than just technical skills but a broad range of advice including how to decide on whether to leave a full-time physical job for online freelancing, navigating the labour platform interface, developing a compelling portfolio, bidding for jobs, identifying reliable from scamming clients, or pricing oneself or a project. Understandably, these kinds of information that a platform worker requires can only be drawn from those who have experienced platform labour directly. As such, YouTube brokers, who are platform workers as well, found themselves in a good position to turn their experience into "expertise videos" that were packaged with the right engagement strategies to suit the needs and aspirations of their community of followers. This is akin to what Burgess and Green (2018) call "vernacular expertise," which pertains to obtaining a critical and literate understanding of the platform's affordances and the attention economy surrounding it to enable a content creator to effectively attune one's content and communication strategies to the social and cultural norms of the platform and the targeted community.

Snippets of training sessions were shared for free but many of the videos were introductions to paid training programs. Posts also included live interviews featuring guest speakers to demonstrate successes in online freelancing, even by new entrants. For example, Luarca (2019) quips on her video, "Learn tips on how to pass and get an insider look in REV! This is an interview featuring one of our viewers (name omitted), who has been with REV since 2016!" Rev is a cloudwork platform similar to Upwork and is starting to become a popular option for aspiring Filipino workers. Another common and primary content on their YouTube videos was about financial sustainability, where brokers coached workers on how to price one's skills, how to increase income streams, and how to invest their earnings given the intermittent availability of work that many freelance workers are concerned about. Their position as "experts" reflects a semblance of organisational hierarchy often seen in traditional work setups. Interestingly however, they balance expertise with relatability by highlighting their belonging to the community, commonly engaging the terms, "we," "us freelancers."

Brokers employed a range of communicative strategies. Highlighting one's personal narratives is affective because this can boost attention and craft a notion of authenticity and relatability for their audiences. This is akin to micro-celebrity strategies of increasing influence and maintaining a "perception of interconnectedness" between the influencer and their followers

(Abidin, 2015, p. 1). The idea that they started as an ordinary digital worker and worked their way up by employing the same methods that they preach in their videos is an important feature of their strategies. The recognition of the coach's "personal story of success" is a demonstration of aspiration and possibility – that through the use of such strategies one can overcome the difficulties of platform labour and actually achieve success. Many of their followers expressed gratitude for "free tips," and in turn, influencers actively acknowledge their subscribers or fans in the form of "shoutouts."

On their videos, brokers asserted their expertise – whether in social media management, podcasting, digital marketing, video production, or virtual assistance – and highlighted their capacity to broker their community of mentees into clinching better projects, higher profile clients, or in choosing the best platform for their skills. The role of brokers in coaching skills or navigating different platforms becomes particularly important given the tendency for many platform workers to move across different kinds of jobs and platforms. To further demonstrate their expertise and care for the community while sharing nuggets of their success, they offered giveaways to mark their online career milestones, such as reaching 300,000 subscribers or gaining six-digit incomes. These online draws or YouTube livestream giveaways were usually sponsored by an industry partner. The giveaways can be in the form of cash sponsored by a financial payment partner such as *PayMaya*, a fintech company, to encourage freelancers to use their online platform or a free laptop sponsored by a computer store. The mechanics they share were often related to subscribing to their social media accounts, subscribing to their channels, liking a Facebook page, and engaging with their existing content. Although they promoted it to give back and celebrate their milestone, it served a dual purpose as they could gain more followers and in turn could monetise these engagements. It illustrates how responsiveness and reciprocity validate their position and that they use their influence in the online freelance community to monetise this labour.

For many brokers, their videos on YouTube served as teasers to paid training packages. Based on our research, the cost of a course could range from ₱3,000 or US$55 for basic digital skills training to ₱15,000 or US$300 for a targeted course on social media marketing or virtual assistance. Based on an interview with a broker, a beginner course on how to become a freelancer costs ₱5,000 (US$100). Although prohibitive for a regular platform worker, some of them would subscribe to such courses, sometimes with borrowed money, with the promise of success made visible through the broker's display of growth in earnings or by featuring mentees or regular subscribers who have become successful. Regular subscribers were invited to share their learning experiences and achievements in the comments and the brokers would often use this as content for succeeding videos.

Community Building and Solidifying Influence: YouTube Brokers as Industry Bridges

Networked connectivity plays a huge role in enabling encounters and interactions among users in a particular platform. Like traditional brokers (Stovel & Shaw, 2012), YouTubers facilitate connections: 1) between them and their followers/subscribers; 2) between workers; 3) between workers and platforms; 4) between potential clients and aspiring workers; and 5) between workers and their digital labour aspirations. Personalised narratives, which follow previous practices of curatorial practices on blogs (Hjorth, 2011), serve as a focal point for imaginaries of entry in the gig economy or for upward mobility. This also highlights aspirational labour that valorises future rewards (Duffy, 2017). However, while narratives as forms of confessions (Strangelove, 2010) highlight their personal journey, these also articulate the presence of help and support from the community.

YouTubers establish connections between themselves and other platform workers by offering a range of advice and tips (Burgess & Green, 2018), and when they have established their expertise they would organise offline and online events and other gatherings to sustain their connection with their viewers and subscribers who now consider themselves as mentees of the broker. In the process of their coaching work, they also build connections between workers of similar experience. All our influencer informants emphasised the importance of online communities to maintain their network of workers and grow their number. Moreover, the influencers also recognised that building a community is the fastest way to grow one's brand organically because it allows the audience to join in the conversation and share content. Notably, a bigger community also equates to higher monetary income. The activation of an "imagined community" (Anderson, 1991, 1983) of online freelancers was a crucial feature of their strategy. Through the use of terms of collective endearment such as addressing their audiences as "*Ka-homemates*" a Filipino colloquial term known as an "online colleague," or "*Ka-OFF*" (fellow Online Filipino Freelancer), they created a sense of common identity for workers who perform different jobs, who may have never met before, and who were physically and geographically distant. Working online can be isolating, and terms such as *Ka-homemates* helped bridge the gap and fostered a type of semblance of interaction where one can learn from a virtual colleague. In another example, a freelancer who branded herself as an advocate of the welfare and career advancement of freelancer moms called her viewers as "*momshees*," a term of endearment implying a sense of belonging with other mothers. This appealed to her targeted demographic of work-from-home moms, but also in this case a playful take of the first syllable of the broker's name Sheena ("shees"). Likewise, although not yet as widely subscribed as other influencer channels, there was "Pinoy Homebased Dads." Led by a freelancer Dad, the influencer hosted "*virtual tagayan*" (online drinking) sessions relatable to Filipino fathers in between discussions on how to boost one's confidence as a freelancer and "masterclasses" on various topics related to online freelancing. A sense of community is a valued commodity in online freelancing because it employs an affective bond to foster a connection with their followers amid varied professional identities, levels of experience, and physical distance.

It is through these imagined communities that they offered a diverse range of coaching resources that allowed them to build reputation and trust. In turn, workers established skills and social capital by being attached to the broker and by gaining a community of support crucial in an ambiguous labour environment. As with traditional brokerage processes (McKay & Perez, 2019), the YouTuber can then monetise the value of their work while workers continually see the need for them due to the perceived value that they can offer. Labour brokers use affective bonds to articulate their expertise and persona in the content they share on YouTube. Crucially, as key movers who promote the viability of platform labour through these bridging strategies, they not only bridge workers with them and with each other, they essentially function as bridges between workers and digital platforms, as well between workers and clients too.

Community building was mobilised by communicating authenticity and relatability. Two points identified by Williams (2016) serve as a guide to examine the construct of authenticity in online influencer cultures: the first is about finding an authentic self, and second is divorcing oneself from an agenda and finding one's authentic voice (p. 127). The first point illustrates how the audience or readers within the online freelance community gravitate to the message that is rooted from personal experience and struggles. Personalised narratives help these brokers appear authentic and allow them to maintain connections and sustain monetisation (Garcia-Rapp, 2017). Authentic communication implies an attempt to balance the potential of monetary gains and conveying one's genuine intent to help, as a freelance influencer explained in an interview:

What people love about me is being genuine and being authentic, I'm not trying to make money every single time like some people say that I could monetise my audience so much and that I could make so much money from the people who follow me.

Of course, this broker also gained financially from the courses he offered, and the availability of free courses served an important purpose of attracting workers and showing a proof of concept. According to our informants, people eventually know if a skill-maker's posts are driven by personal agenda and the workers get turned off by this. People's comments about doubting the authenticity of an influencer can "infect" others in the community and create ripple effects to the rest of other subscribers. For instance, although an influencer can earn from advertisements in their YouTube channel, engaging too much in sponsored posts can backfire and drive followers away. Connected to authenticity is responsiveness and reciprocity. According to one broker we interviewed, "to be authentic is to be responsive and to engage with other freelancers' issues, concerns, and stories." A tough effort, digital labour brokers on YouTube were compelled to address inquiries attentively on chats and comment sections or in succeeding videos and lives, crucial in maintaining their authenticity and credibility.

Finally, although visibility is crucial in sustaining a community, influencers appear to also work on "calibrated visibility" (Abidin, 2015, p. 4; Soriano & Panaligan, 2019). Our interviews support the idea that being consistent and visible in the community should be calibrated, where brokers balance between the constant production of valuable content and simultaneously allowing community members to engage organically such that skill-makers do not dominate the conversation. To avoid the impression of merely taking advantage of their subscribers, brokers worked to maintain the right balance between promotional vis-à-vis non-sponsored videos and requesting for subscription and likes while supplying useful content for the community (Soriano & Panaligan, 2019). To know what works, our interviews showed that digital labour influencers also carefully studied the pattern of interaction of their community-members.

Maintaining a sense of community through everyday interactions, they would regularly invite viewers to subscribe, join contests, interact with other followers, and share their own experience. For example, some influencers would launch contests such as *"Find the Hidden Code Contest: Whoever finds the code hidden somewhere in this video and comments it first in our comment section below wins P50 load!"* – these kinds of activities were common and functioned well to sustain the attention of their subscribers while expanding the viewership and engagement with their videos through the promise of an immediate reward.

The brokers not only showcased their own success stories, but they also highlighted the successes of others' belonging to their community of "mentees" or subscribers, and occasionally, the achievements of other influencers. Showcasing the success of their mentees allowed them to demonstrate social proof of their tips and capacity to mentor – and having this conveyed by others in the community, such as those who applied the coach's lessons to reach the path to success, lends it more believability. Brokers explained that this social proof is valuable because many freelancers need to battle with questions about the viability and sustainability of platform labour by friends or family members. Often, freelance workers were uncertain whether earning online is truly possible given the widespread cases of scams. On the other hand, influencers like Ruiz (2019) also cited and praised the success of other brokers, and this demonstrated their positivity and congeniality as "authentic" and sincere advocates of the freelancing community.

Given the above, it would appear that the YouTubers' brokerage role lies not only in putting together disparate information needed by aspiring platform workers, but also by bridging geographically isolated workers through the creation of a parasocial community on YouTube. Yet, ultimately, brokers perform the recruitment role for labour platforms, albeit informally and indirectly. As key movers inspiring local workers to take up platform labour, to training them as entrants to this industry, brokers bridge workers with platforms by making the ambiguous digital platform environment less complex for local workers – a gain for digital labour platforms. Indirectly, they also serve as promoters of the government's vision of Filipinos becoming "world-class workers."

Constructing Norms and Value Standards in Platform Work

Beyond cascading strategies and imaginaries of success, brokers constructed standards of value and norms in the industry. The construction of value standards pertains to setting norms of what counts as "success," what counts as a "good client," what constitutes a "good digital worker," a "good project," or even a "conducive workspace." Although not well-emphasised in the literature, traditional brokers do this too by advising their clients on which employers or potential employees can be "trusted," what kinds of jobs are "best suited" to one's skills, or what investments can give the client the "best value."

Notably, the brokers' labour on YouTube nudged workers to rethink their standards of "good work," showing that digital labour can be a site for negotiating flexibility and the attractive fluidity of the spaces of work-and-home. Beyond this spatial flexibility that allows them to be connected to the infrastructures of global capital while avoiding infrastructural challenges of heavy traffic, bad roads, or inefficient public transport, the brokers also highlighted flexibility in terms of limitless social mobility – but not without embodying an entrepreneurial and resilient character (Soriano, 2021). Concretely, brokers showed that "the more one works hard, the greater the success." These aspirations would not make much sense to workers amid the ambiguities tied to platform labour. The work of brokers became crucial in illustrating the opportunities while presenting entrepreneurial strategies and vignettes of success that will help workers visualise that the possibilities may indeed be actualised. The framing and visibilising of success stories as attainable through dedication and hard work also functioned to dictate images and norms of a good worker and success in this industry.

Yet, as helpful as the sharing of personal strategies could be, the narratives downplayed the challenges that workers experience, although some of the workers would occasionally share their gripes about platforms or clients in these spaces, too. In response, the coaches would often emphasise the importance of hard work, resilience, and that digital labour "success" could be achieved with the "right frame of mind," as well as an ample dose of "grit and guts." It is important to emphasise that over time, other members of these communities were observed to take on the role of coaches for one another, and this is done by posting their own "winning strategies" and achievements in the comments, again presenting an image of possibility, but also their own version of success that aspiring freelancers can emulate. The visual nature of YouTube interactions made such messaging of success possibilities more affective and potent. Yet, this created lofty expectations that everyone can and should be like them, successful, entrepreneurial, and well-networked.

YouTube as a Site for Platform Labour Brokerage

In this chapter, we examined the conditions that give rise to digital labour brokers, their role in the local platform economy, and how through their communicative strategies they sell not only skills but also digital labour aspirations while also building a community among online freelancers and setting norms in this largely unregulated sector. This brokerage work enabled them to benefit economically and socially in the process. In turn, we also considered how labour mobility or precarity was made possible and organised through the work of these actors.

We argue that selling aspiration in the context of platform labour involves providing a vignette of possibility coupled with know-how on how to achieve this possibility. YouTube affords the display of these narratives of possibility visually and socially, constructing the right affect to persuade an aspiring worker to consider platform labour. The aspirational aspects of platform labour – obtaining competitive jobs without formal university degrees, gaining influence despite one's disability, nurturing the family while becoming economically productive, or avoiding the horrible traffic conditions of travel to and from a workplace – are already well-promoted by the government, along with these brokers. Platform labour is shaped by workers' imaginaries of their capacity to overcome the difficulties inherent in digital labour and, consequently, to attain the promise of a successful future. This includes building a resilience-based reputation and, subsequently, visualising success in terms of what they have managed to achieve or buy that results from their work, all actively shared by brokers and their community of subscribers on YouTube. As we have presented in Chapter 2, this practice of building a reputation around resilience in the context of platform labour has strong colonial inflections as it reinforces a dynamic that makes workers feel that their drive for self-improvement and social mobility are in line with what is deemed desirable by global capital. In turn, this "resilient" mindset may also make them susceptible to exploitation and self-exploitation, leading them to put in extra hours imagining that future returns from their hard work will come by in the future.

Importantly, platform labour brokers on YouTube, by selling skills and aspirations and forging communities among Filipino online freelancers, play a symbolic role in mediating the relationship between the local workforce and digital labour platforms. Despite their influence in the sector, which is likely to grow given the continued popularity of platform labour in the country, their brokerage work is largely unregulated by formal state institutions nor recognised by labour platforms. Yet, by brokering aspiration and possibility through a combination of affective and community-building strategies, they essentially play the role of promoters and matchmakers between labour platforms, foreign clients, and Filipino workers.

YouTube plays a critical role in the production and circulation of personalised content, while the aesthetics, narratives, and discursive styles of these influencers convey aspirations of economic mobility for viewers, community members, or aspiring platform workers. Its multiple affordances allow these brokers to thrive, visibilise and amplify their content, facilitate engagement, and expand their network of followers and other influencers. This interaction between platforms, influencers, and their viewers engender social support and networking opportunities needed to thrive in the platform economy. Further, brokers build on their branding strategies by making their experiences, struggles, and successes visible, which is crucial in attracting viewers to imagine and aspire for possibilities for work in the digital economy, figure out solutions to platform labour ambiguities, or to feel affirmed about their work choices. Stories of success and sharing of tactics in navigating a new digital market become essential in supporting individuals with shared experience in the struggle to actualise their aspirations within a precarious labour environment.

In this chapter, we have shown how platform workers' dynamic negotiation of agency enables workers to seek a certain relief from the structural constraints and precariousness of platform labour while allowing them to envision or partially (or fully) achieve their aspirations, through the work of brokers. In turn, brokers obtain patronage capital by obtaining influence as industry leaders. While the more enterprising followers can emulate them, others continually cling to the celebratory promises of platform labour and rely on the inspiration and social capital gained from being associated with brokers, valorising them even when the promises are not yet exactly realised. It is in the heart of these contradictions made palpable by labour brokerage that platform workers envision success in the digital economy despite uncertainty on whether these can in fact be actualised.

Platform labour brokers have strong *group orientation bias* given that they are viewed as members of the same community (Stovel & Shaw, 2012; Gould & Fernandez, 1989). Thus, it is in presenting their capacity to bridge the difficult realities of platform labour with tangible strategies and affective success vignettes that these skill-makers create value for themselves and their work as brokers. Yet, it is also important to emphasise that these operate largely to sustain, rather than challenge, the power embedded in labour platforms and therefore work to reinforce and even strengthen the platforms' capitalistic motives. Thus, while the bias and orientation of brokers may appear to be towards the workers, a closer analysis would show that the performance of their brokerage role ultimately promotes the ideological discourse of labour platformisation, performing the indirect role of recruiters and promoters (Soriano, 2021). While they play a crucial role in helping workers make sense of platform labour when they have no better alternatives nor access to similar forms of dynamic guidance elsewhere, they also cascade imaginaries (and sometimes possibilities) that workers hang on to as these nurture their economic, social, and cultural aspirations. But precisely because they are dependent on platforms for their survival, and by instigating the very platform logics, their brokerage function does not include, for example, mediating for workers to bargain – with the government or with platforms – for systematic changes that can facilitate equitable pay or better labour arrangements for workers. More recently, we notice the emergence of new brokers who coach workers on how to "break free" from the regular platforms and become truly "independent" by boosting their profiles and portfolios in LinkedIn or Facebook. It is an interesting development and direction to follow as we continually observe the role of social media influencers in brokering in the platform economy.

Platform labour brokerage presents the duality and the complexity of influence, intermediation, and survival in the local platform economy. This chapter has presented the contingent and fluid forms of labour brokerage strategies that unfold on YouTube, pushed by techno-capitalist logics, state shortcomings, locally entrenched celebrity and patronage culture, experimental possibilities, and entrepreneurial logics – and which yield critical insight into the facilitation and conditioning of digital labour in the Global South.

References

Abidin, C. (2015). Communicative ♥ intimacies: Influencers and perceived interconnectedness. *Ada: A Journal of Gender, New Media, and Technology, 8*, 1–16. doi:10.7264/N3MW2FFG

Abidin, C. (2017). #familygoals: Family influencers, calibrated amateurism, and justifying young digital labour. *Social Media + Society, 3*, 1–15. doi:10.1177/2056305117707191

Anderson, Benedict. 1991 [1983]. *Imagined communities: Reflections on the origins and spread of nationalism.* Verso.

Baym, N.K. (2010). *Personal connections in the digital age* (First ed.). Polity Press.

Burgess, J., & Green, J. (2018). *YouTube: Online video and participatory culture* (Second ed.). Polity Press.

de Peuter, G., Cohen, N.S., & Saraco, F. (2017). The ambivalence of coworking: On the politics of an emerging work practice. *European Journal of Cultural Studies, 20*(6), 687–706. doi:10.1177/1367549417732997

Duffy, B.E. (2017). *(Not) getting paid to do what you love: Gender, social media, and aspirational work.* Yale University Press. https://www.jstor.org/stable/j.ctt1q31skt

Florida, R. (2014). *The rise of the creative class – revisited: Revised and expanded.* Basic Books.

Garcia-Rapp, F. (2017). Popularity markers on YouTube's attention economy: The case of Bubzbeauty. *Celebrity Studies, 8*(2), 228–245. doi:10.1080/19392397.2016.1242430

Graham, M., Hjorth, I. & Lehdonvirta, V. (2017). Digital labour and development: Impacts of global digital labour platforms and the gig economy on worker livelihoods. *Transfer: European Review of Labour and Research, 23*(2), 135–162. doi:10.1177/1024258916687250

Hesmondhalgh, D. & Baker, S. (2010). 'A very complicated version of freedom': Conditions and experiences of creative labour in three cultural industries. *Poetics, 38*(1), 4–20. doi:10.1016/j. poetic.2009.10.001

Hjorth, L. (2011). It's complicated: A case study of personalisation in an age of social and mobile media. *Communication, Politics & Culture, 44*(1), 45–59. https://search.informit.org/doi/10.3316/informit.127649781513814

IBON Foundation. (2017). Contractualization prevails. *IBON Facts and Figures, 40*(7). IBON Foundation.

Kapunan, R.P. (1991). Labour-only contractors: New generation of 'Cabos.' *Philippine Law Journal, 65*(5), 326.

Kelly, P.F. (2001). The political economy of local labour control in the Philippines. *Economic Geography 77*(1), 1–22. doi:10.2307/3594084

Lange, P.G. (2019). *Thanks for watching: An anthropological study of video sharing on YouTube.* University Press of Colorado.

Lehdonvirta, V. (2016). Algorithms that divide and unite: Delocalisation, identity and collective action in microwork. In J. Flecker (Ed.), *Space, place and global digital work* (pp. 53–80). Palgrave Macmillan.

Lin, W., Lindquist, J., Xiang, B. & Yeoh, B.S.A. (2017). Migration infrastructures and the production of migrant mobilities. *Mobilities, 12*(2), 167–174. doi:10.1080/17450101.2017.1292770

Lindquist, J., Xiang, B. & Yeoh, B.S.A. (2012). Opening the black box of migration: Brokers, the organization of transnational mobility and the changing political economy in Asia. *Pacific Affairs 85*(1), 7–19. doi:10.5509/20128517

Marwick, A.E. (2013). *Status update: Celebrity, publicity, and branding in the social media age.* Yale University Press.

McKay, D., & Perez, P. (2019). Citizen aid, social media and brokerage after disaster. *Third World Quarterly, 40*(10), 1903–1920. doi:10.1080/01436597.2019.1634470

McKenzie, M.J. (2020). Micro-assets and portfolio management in the new platform economy. *Distinktion: Journal of Social Theory.* doi:10.1080/1600910X.2020.1734847

Mimi Luarca. (2019). *Be a part-time transcriptionist in REV and earn 11,000 pesos! Work from home English subtitles.* [Video]. Retrieved 12 September 2020, https://www.YouTube.com/watch?v=V6oESxIwowM

Neilson, B., & Rossiter, N. (2008). Precarity as political concept, or, Fordism as exception. *Theory, Culture & Society 25*(7–8), 51–72. doi:10.1177/0263276408097796

Ofreneo, R.E. (2013). Precarious Philippines: Expanding informal sector, 'flexibilizing' labour market. *American Behavioral Scientist 57*(4), 420–433. doi:10.1177/0002764212466237

Padios, J.M. (2018). *A nation on the line: Call centers as postcolonial predicaments in the Philippines.* Duke University Press.

Payoneer (2020). *Freelancing in 2020: An abundance of opportunities.* Retrieved 10 July 2021, https://pubs.payoneer.com/docs/2020-gig-economy-index.pdf

Rodriguez, R.M. (2010). *Migrants for export: How the Philippine state brokers to the world.* The University of Minnesota Press.

Ruiz, J.T. (2019, November 27). *Online jobs at home Philippines for beginners (full tutorial).* [Video]. YouTube. https://www.YouTube.com/watch?v=GUo_3FZtISA

Santos, S. (2019, Aug 12). How to make money on Fiverr without skills/ How to make money on Fiverr for beginners. [Video]. YouTube. https://www.youtube.com/watch?v=0tQ2I4WhSCw

Senft, T.M. (2013). Microcelebrity and the branded self. In J. Hartley, J. Burgess, & A. Bruns (Eds.), *A Companion to New Media Dynamics* (pp. 346–354). Wiley-Blackwell Publishing.

Shresta, T., & Yeoh, B.S.A. (2018). Introduction: Practices of brokerage and the making of migration infrastructures in Asia. *Pacific Affairs 91*(4), 663–672. doi:10.5509/2018914663

Silarde, V.Q. (2020). Historical roots and prospects of ending precarious employment in the Philippines. *Labour and Society, 23*(4), 461–484. doi:10.1111/wusa.12484

Soriano, C. R. (2021). Digital labour in the Philippines: Emerging forms of brokerage. *Media International Australia*. doi:10.1177/1329878X21993114

Soriano, C.R., & Cabañes J.V. (2020a). Entrepreneurial solidarities: Social media collectives and Filipino digital platform workers. *Social Media + Society, 6*(2),. doi:10.1177/2056305120926484

Soriano C.R., & Cabañes J.V. (2020b). Between 'world class work' and 'proletarianised labour': Digital labour imaginaries in the Global South. In E. Polson, L. Schofield-Clarke, & R. Gajjala (Eds.), *The Routledge Companion to Media and Class* (pp. 213–226). Routledge.

Soriano, C.R., & Panaligan J.H. (2019). Skill-makers in the platform economy: Transacting digital labour. In A. Athique, & E. Baulch (Eds.), *Digital transactions in Asia: Economic, informational and social exchanges* (172–191). Routledge. doi:10.33767/osf.io/z4wun

Strangelove, M. (2010). *Watching YouTube: Extraordinary videos by ordinary people*. University of Toronto Press.

Williams, R. (2016). *The influencer economy: How to launch your idea, share it with the world, and thrive in a digital age*. Ryno Lab.

Wood, A.J., Lehdonvirta, V., & Graham, M. (2018). Workers of the Internet unite? Online freelancer organisation among remote gig economy workers in six Asian and African countries. *New Technology, Work and Employment, 33*(2), 95–112. doi:10.1111/ntwe.12112

6   Politics

Abstract

This Chapter focuses on the brokering of Philippine politics on YouTube. We zoom in on how political influencers, well-versed with platform vernaculars, affordances, and porous governance structures, create content that convey hyper-partisan political narratives advancing a historical revisionist and political agenda. The chapter showcases how Filipino YouTubers mobilise affective aspirations on the nostalgia of a progressive state during the regime of the late dictator and former President Ferdinand Marcos, while dismissing the documented violent atrocities surrounding that political regime thereby facilitating political discourses in favor of the reinstatement of another Marcos to the Presidency. By purporting to be the "voices from the margins" and facilitating "illusions of discovery," the parasocial relationships developed by YouTubers along with the platform's participatory and "epistemic cultures" gives them the legitimacy to construct their version of political history while brokering a connection between YouTube voter publics and political actors.

Keywords: Philippine politics, political brokerage, historical revisionism, Martial Law, Marcos, YouTube, disinformation

Our interest in the role of YouTube for politics began with a Grab ride around the city of Manila in 2019. Unlike cab drivers of the past who would be tuned in to AM or FM radio stations while they drove for passengers for the day, this Grab driver, *Kuya* Armand (not his real name), is one of the increasing group of drivers who tune in to YouTube as their ambient source of news, entertainment, and political commentary. With free data connection accompanying his Grab service, Armand told one of the authors, Cheryll, that YouTube has become a valuable source of information for him. He tunes in to YouTube regularly, and in particular, to channels that discuss politics. During the trip, the driver would consistently echo the words of a YouTuber, as if to convince his passenger – *"Di ba, si Marcos talaga ang pinakamahusay na Pangulo ng Pilipinas"* (See, Marcos really is the best President the country

Soriano, Cheryll Ruth and Earvin Charles Cabalquinto: *Philippine Digital Cultures: Brokerage Dynamics on YouTube*. Amsterdam: Amsterdam University Press, 2022
DOI: 10.5117/9789463722445_CH06

ever had). While waiting for a green light on Gil Puyat Street, he expressed, *"Ma'am, kung hindi po sana napaalis si Marcos, siguro po mayaman na bansa na tayo ngayon ano?"* (Ma'am, if only Marcos was not ousted from power, maybe we would have been a progressive nation, right?). Throughout the ride, the driver would laugh at the YouTuber's jokes, sometimes shook his head to express disgust over certain topics being discussed, or nod in complete agreement. Just before approaching the destination, he enthusiastically quipped again, *"Mabuting tao talaga si Madam Imelda at napakaganda pa. Dapat kasi si Madam Imelda na lang ang sumunod kay Marcos dahil ibabalik nya para sa atin ang Marcos gold"* (Imelda Marcos is really a kind-hearted and beautiful person. Marcos should have just been succeeded by Imelda, his wife, so that she could have brought back the Marcos gold to benefit all of us). Before alighting, Cheryll was compelled to ask, Kuya (*Brother*), *"ano po ang YouTube channel na yan?"* (what YouTube channel is that?) He replied: *"Si SangkayJanJan yan mam! Subscribe po kayo kasi magaling sya sa history at pulitika."* (That is SangkayJanJan, Ma'am. Please subscribe to his channel because he is really good in history and politics!).

What Kuya Armand shared with us echoed two other past encounters with drivers who were tuned in to YouTube channels for political information in the everyday. But only until meeting Kuya Armand and how he actively echoed the media content that he was consuming did the political power of such YouTubers and their videos strike us. YouTube is becoming a crucial site for political engagement by hosting the savvy work of political intermediaries to advance a politicised agenda. They are like smooth operators who are achieving influence from one subscriber to the next, hooking the ordinary viewer through a potent mix of entertainment, DIY historical analysis, and political commentary. Importantly, their work is inscribed in transformations in "epistemic cultures" facilitated by digital participatory cultures.

In this chapter, we examine the brokerage of politics on YouTube by focusing on the case of historical revisionist discourse on the late Philippine President Ferdinand Marcos and his authoritarian regime. Due to the active contestation of Marcos loyalists of institution-sanctioned historical narratives, historical information about Marcos and his administration remains floating, contested, and unsettled (Reyes & Jose, 2012–2013; Aguilar, 2019, p. 3). A complicating factor is the fact that pertinent scholarly works on history are often inaccessibly written and not easily available to the general public. Since Marcos' death, there have been ongoing attempts to reconstruct his legacy from the atrocities of Martial Law, ill-gotten wealth, cronyism, and economic crisis (Bautista, 2019; Reyes, 2018; Reyes & Jose, 2012–2013; Reyes & Ariate, 2020). Marcos' reputation as a dictator is well-documented, known to be the only President so far to have enacted Martial Law. Martial Law was not just an invocation of the late dictator's emergency powers under the Constitution. Marcos used this power to assume all governing powers, excluded civilian courts, imposed government control over the media, and systematically created a new Constitution "that placed all abuse of power under a cloak of legality for his own ends" (Official Gazette, 2015, n.p.). This suspension of civil law and *habeas corpus* from 1972 to 1981 subjected the Filipino people to military justice, arbitrary detention, torture, and forced disappearances while allowing the Marcos family to accumulate unexplained wealth (McKay, 2020). It is important to note that despite significant documentation connecting the Marcos' to massive corruption and human rights violations (Official Gazette, 2015; Mijares, 1976), the Marcos family continue to be political power holders, having been elected to important national and local positions for the last 20 years and with Marcos' son, Ferdinand Jr., running for the Vice Presidency in the 2016 elections. The Marcos family have not admitted nor apologised for documented ill-gotten wealth, as well as violent abuses during Martial Law, dismissing these as mere tactics of the opposition to malign their name (Reyes, 2018). When we started researching for this chapter, there were speculations that Marcos' son, Bongbong Marcos, was intending to run as President of the country for the May 2022 elections, also as evidenced by the rise of videos supporting his candidacy on various social media channels. He indeed ran for the Presidential post, topping pre-election survey results. As we finalize this book, provisional election results show Bongbong Marcos winning by a significant margin, garnering over thirty million votes or around 56% of total votes (GMA News Online, 2021).

The failure of the succeeding reformist presidencies to instigate meaningful economic and social progress (Webb & Curato, 2019), along with the election of President Rodrigo Duterte who has expressed support for the Marcos family (McKay, 2020), further created a political atmosphere for revisionist narratives to thrive. Duterte's support for the Marcos family, manifested by allowing the burial of the former dictator in the *Libingan ng mga Bayani* (Cemetery of the Heroes) amid the vehement objection of human rights activists, created greater openings for the family to reactivate its political power base by seeking broader national credibility. Further, the reclaiming of democracy in 1986 via a People Power revolution that ousted Marcos has created expectations that development will trickle down for all. Yet, despite the popularity of this democratic revolution that created openings of popular mobilisations and space for dissent not just in the country but in the world, the succeeding Philippine governments failed to facilitate meaningful and equitable development. Progress had been continually slanted to benefit the rich and the powerful classes, consequently facilitating the continued poverty and frustration of sectors that have already suffered during the Marcos regime. These political, economic and social conditions are conducive to historical and political distortion especially with the majority of the population of younger generations having no direct experience or personal memory of the Marcos era.

Pro-Marcos sentiments have been in public debate since the early 2000s through the self-publication of books and manuscripts by authors traced to be cronies and allies under the patronage of the Marcos family (Reyes & Ariate, 2020). However, these were not repackaged for wider mass consumption "until social media became a well-established campaign tool" (McKay, 2020). Amid a strong opposition to a renewed Marcos rule, propagandists require intermediaries on social media to help drive a preferred political historical narrative. This chapter focuses on the work of these political intermediaries or brokers on YouTube, who have begun to gain significant following and subscribership of their channels.

YouTube and Political Engagement

There has been optimism that social media would foster digital and participatory democracy (Loader & Mercea, 2011). With social media playing a key role in movements around the world in the late 2000s and early 2010s, social media appeared to be the technology critical for toppling down authoritarian regimes (Diamond, 2010, pp 70–71; Howard et al., 2011) as well as capitalistic orders in Western democracies (DeLuca et al., 2012). Social media's role in shaping the public agenda also heightened the expectation of its democratising possibilities, shifting the power from traditional sources and regulators of knowledge and information to ordinary people.

YouTube is an important platform for examining social media's role in politics. The literature on YouTube and political engagement has been continually growing, with analyses on how the platform is utilised for political campaigning and propaganda, political education, as well as activism and organising from below. As presented in Chapter 1, the tremendous uptake and popularity of the platform among users in the Global South, including the Philippines, drives the increasing and creative use of YouTube by political actors for reaching voters and constituents. Filipinos' love for both entertainment and political engagement (A. Pertierra, 2021) also makes them a highly suitable audience for the kind of political content rendered in infotainment style. Yet, the diverse ways political communication is enacted on YouTube can be further explored. As we will point out in this chapter, there are particular affordances of the platform that allow for creative political communication (including manipulation and distortion) facilitated by intermediaries to thrive and get popularised yet remaining obscure from scrutiny.

Although several works examine YouTube's role in facilitating political communication and engagement there is continuing scholarly discussion on what makes YouTube unique for political communication. For example, Gueorguieva (2008) argued that the potential of campaigns to access voters through YouTube is unlimited. How YouTube influenced electoral processes has also been widely studied (for example, Hanson et al., 2011; Towner & Dulio, 2011), although then YouTube was seen to privilege traditional media sources and actors. More recent studies examined how YouTube allowed non-conventional actors to drive political discussions (Dylko et al., 2012), and even previously unknown people to run for elections (Chang & Anh, 2020). New formats of political advertising by non-traditional actors were also found to receive more attention from the audiences (Ridout et al., 2010). Broadly, these works characterise YouTube as an emerging political sphere that blends propaganda and political persuasion, celebrity culture, and local political cultures through political actors and their intermediaries.

The YouTube platform has several embedded features to boost social interaction (Benevenuto et al., 2008) such as the YouTuber's capacity to integrate creative visual modalities to facilitate political persuasion, the users' ability to subscribe, like/dislike, or comment on a video, or share a video to other social network platforms (Feroz Khan & Vong, 2014, p. 631). This is supported by the fact that the evolution of YouTube through the introduction of paid advertising is shifting academic analysis on YouTube from the viewpoint of participatory culture towards an analysis of a "hybrid cultural–commercial space" (Arthurs, Drakopoulou, & Gandini, 2018; Lobato, 2016, p. 357) – where various forms of political content become entangled with commercial or entertainment styles of presentation. Scholars and the media have directed the spotlight on the role of YouTube as a "political influencer" (Lewis, 2019; Sucio, 2019) in the international and local political arena, and to an extreme extent, even as a driver of "political radicalization" (Fisher & Taub, 2019; Dagle & Fallorina, 2019) due to its possibilities for facilitating socially immersive engagement. This is, in part, enabled by its platform affordances and logics that facilitate the creation, curation and engagement of political content, produced and participated by YouTube users and content creators (Chang & Anh, 2020). Among them are micro-celebrities

(Marwick, 2015) as emerging political actors in the platform, who build their own networks of followers by capitalising on YouTube's creative and participatory culture, while inserting snippets of politicised information or opinion. The interplay of visual and auditory styles contributes positively to the "virality" of a video (Benevenuto et al., 2008; Kaplan & Haenlein, 2010) that can then trigger political engagement in the context of platform convergence. Broadly, these works argue that comment cultures on political videos help sustain civic cultures that allow for both divisive and inclusive political debate.

Philippine Politics and Brokerage on YouTube

This chapter examines political brokerage on YouTube by looking into the construction of political discourses and the action of a genre of political actors whom we call brokers to advance a political agenda through historical revisionism. We will show that YouTube is an important socio-technical driver of a political brokerage process as it allows emerging influencers to bridge politics, history, and local audiences while being able to advance a monetary and political agenda. The brokers' strategic use of the platform privileges them with visibility and renders relevance to their narratives that together curate political and historical information in this contemporary political scene.

The enactment of politics in the Philippines is underscored by brokerage. As we discussed in Chapter 2, brokerage works well within pervasive "patron-clientilistic relations" (Nowak & Snyder, 1974; Scott, 1972) that underscore how brokers obtain legitimacy to bridge ordinary constituents and political elites; in turn, elites benefit from this brokerage process by retaining their control over constituents (Nowak & Snyder, 1974, p. 23). Patronage structures are deeply embedded and have become long standing features in Philippine society (Kerkvliet, 1995). For example, politicians can appear distant from the masses and need political brokers, often emerging from the middle classes, who can intermediate between them and the masses (Nowak and Snyder, 1974). Here, the brokerage function is based on an exchange – the broker promises resources or an illusion of closeness to the politician, in exchange for a vote. Aspinall and colleagues (2016, p. 194) also emphasised the role of the *liders* (campaign staff and vote brokers) in the contemporary Philippine political structure, who function to enliven political campaigns through in-person encounters that bring political candidates (and political aspirations) closer to the public through "personalised patronage." It is through personalised patronage that these brokers obtain influence by directly mobilising people's interests and support, making them crucial players in the political process. Where politics is about mobilising interests, persons able to influence and control the flow of information and resources set the agenda and obtain power in shaping political outcomes. As the political economy has become differentiated and with politics becoming more embedded in social media, specialised patron-client relationships have emerged in new forms such as political machines and social media architects (J. Ong & Cabañes, 2018; Ong, Tapsell, & Curato, 2019; Ressa, 2016). However, the style and character of social media brokerage remains to be fully studied and unpacked along with platform affordances.

Methodology

Platforms like YouTube not only mediate public discourse, but "call it into being, give it shape and affirm its basic legitimacy" (Gillespie, 2018, p. 20). Through its affordances, constraints and logics, YouTube structures constitute the nature of interactions and transactions brokered in the platform (van Dijck & Poell, 2013). By connecting political actors, cultural workers and publics, commodifying their interactions, and profiting from their transactions (Stovel & Shaw, 2012), YouTube acts as a socio-technical broker involved in the larger process of brokerage. Political brokerage on YouTube not only happens *through* the platform but *with* the platform. To capture this process, the research adopted Rogers' (2013) methodological approach of "following the medium" and navigated political brokerage on YouTube through the distinct ontological contours of the platform. This means performing the "native" logics of the platform, from searching, viewing, to interacting with the interface where political interests are brokered (Airoldi et al., 2016).

We reproduced "native" user transactions on YouTube to surface, identify and analyse pro-Marcos videos and their broker (Bucher, 2016; Diakopoulos, 2015; Sandvig et al., 2014). The first step is determining the "native" user behaviour within the YouTube platform, which is to search using keywords to discover relevant videos about the political discourse around Marcos and watch the top search results and the succeeding recommended videos in the "Up Next" column in the YouTube interface. Using the Google Trends tool, we identified the keywords "Marcos History" as the top breakout query (indicated by a sharp increase in the volume of search) on YouTube related to Ferdinand Marcos from October 2015, when son Bongbong Marcos joined the vice presidential race for the 2016 elections, to March 2020. The second step is re-enacting this native YouTube behaviour using a clean, cookie-free browser, and search for the identified keywords. From the search results, we documented and watched the top videos that fit the profile of a broker – informal, "amateur" content creators (Lobato, Thomas & Hunter, 2012) unaffiliated with professional institutions or groups. The third step involved collecting the top amateur suggested videos appearing on the right-hand sidebar for each viewed video. This process was iterated three times within a week by two researchers and generated a list of about 300 videos, with a number of videos appearing multiple times.

From the video database, we shortlisted visible videos and identified the brokers who produced them. We further investigated each channel if they have published more Marcos-related videos that did not appear in our initial data collection process. The brokers for analysis were narrowed down to those with at least 1,000 subscribers, which is the minimum subscriber count to start monetising a YouTube channel, and at least have produced five pro-Marcos videos. We analysed their videos through close reading and supplemented the data with a general examination of the brokers' channels. The succeeding discussion presents the characteristics of these brokers, their channels and their videos, the aspirations they promote, as well as their discursive styles and platform-relevant strategies. Selected quotes were translated from Filipino or a mix of English and Filipino to English.

Political Brokerage on YouTube

Brokers, videos, and channels

For this chapter, we analysed eight brokers with at least five pro-Marcos videos (following the title/description of the video), and with a range of 2,000 to close to 400,000 subscribers. Despite some brokers with relatively less subscribers, the videos they produce can gain 50,000 to 850,000 views each, indicating that their influence as political brokers might be contingent on a number of factors. Majority of the brokers in our corpus joined YouTube in 2016 and their channels published after the 2016 national elections, except for one, which was created in 2015 (Please refer to Table 6.1).

A few salient characteristics were notable among the brokers. All brokers appeared to be early to middle-aged males (at least in their self-presentation or persona). Prominent among the brokers was the use of pseudonyms (i.e., *Sangkay Janjan, OHJAYCEE, Jevara PH*) and other key "self-branding"

strategies such as the use of a logo and video branding styles in the opening
and closing credits of their videos, akin to the branding work of broadcast
networks that utilise slogans and logos to convey their brand. The notion of
self-branding implies that micro-celebrities need to differentiate themselves
from the rest by creating a *persona,* a character or role created to project
themselves in the content or product they create, and eventually display
influence over people (Abidin, 2017). Consistent with the work of micro-
celebrities, visual and audio styles of self-branding included the consistent
use of colour scheme, font, layout, music bed. This creation of a "persona"
illustrates how brokerage involves the "marketing of personal sensations"
through the use of a personal touch in order to attract and sustain an audience, while at the same time differentiating one's "goods" from the rest.

Consistent to all brokers was their anti-establishment rhetoric, particularly the rejection of traditional sources of knowledge and information
(e.g., educational and historical institutions, scholarly work, mainstream
media). Common to all channels, either expressed overtly in the channel
description or mentioned in their content, was how they tackled Philippine
"history" from their own perspective and research, implying that what they
present may conflict with "official accounts" from historical records and
scholarly works. One such broker, *Filipino Future,* described the channel
as "about Philippine History, from a perspective of a patriotic Filipino, a
Marcos Loyalist and supporter of nationalistic government." In another,
Sangkay Janjan's' videos opened with "I would like to reveal the truth about
the history of our nation from the perspective of us millennials." These
openings implied that their versions of history have more credence than
what is written in history books. This appears not only as a strategy for
promoting legitimacy but as part of their branding strategy as the alternate,
if not the more legitimate epistemic source of truth. The caveats expressed
in these channels, together with their anti-establishment rhetoric, establish
the idea that they are one with their audiences, the ordinary Filipino, who
should go beyond the controls of traditional gatekeepers of knowledge and
seek the "truth" about history on their own.

Whether to establish authenticity or to facilitate financial sustainability,
brokers' channels were notably characterised by a mix of political and
non-political content (i.e., sharing of lifestyle and survival tips, travel, social
commentary, as well as conspiracy content), and where the content was
strategically placed in thematic brackets. This strategy worked to effectively
balance between performativity and authenticity, while at the same time
sustaining constant community engagement and monetisation of content.
It was also common for the videos in these channels to ride on trending

Table 6.1　Political brokers and sample of videos analysed

Broker	Date of joining YouTube	Subscribers	Number of videos	Sample video analyzed	Date published	Views
Sangkay Janjan TV	Dec 1, 2017	702,000	826	KUNG NATULOY ANG PLANO NI MARCOS, HIGIT PA SA AMERIKA ANG PILIPINAS *(If Marcos' plans pushed through, the Philippines would be better than the US)*	July 1, 2019	1.8M
OHJAYCEE	Dec 8, 2018	11,300	383	Ang mga HULING SALITA ni Ferdinand Marcos *(Ferdinand Marcos' last words)*	Jan 21, 2020	1.069M
Jevara PH	Nov 14, 2017	844,000	349	Mga Magandang Nagawa ni Ferdinand Marcos | Jevara PH *(The great accomplishments of Ferdinand Marcos)*	Sept 12, 2019	508,000
Tinig PH	Sept 20, 2018	212,000	409	Pinoy Trivia "Alam mo ba?" Paano Namatay Ang Mga Pangulo Ng Pilipinas *("Did you know?" How the Philippine presidents died)*	Feb 13, 2020	450,000
Filipino Future	Nov 26, 2017	151,000	213	Ang Totoong Dahilan Bakit Ayaw Nila Maging Presidente Si Bongbong Marcos *(The real reason why they don't want Bongbong Marcos to be president)*	April 7, 2020	1.76M

Broker	Date of joining YouTube	Subscribers	Number of videos	Sample video analyzed	Date published	Views
Birador HQ (recently changed to Chopseuy King)	Nov 26, 2017	218,000	315	ANG MGA NAGAWA NI PNOY SA ATING BANSA, MAS TAMANG TANUNG PALA EH MAY NAGAWA NGA BA SYA? #birador (What were the accomplishments of PNoy (former president), or is the better question is, did he do anything?)	Mar 14, 2020	377,000
Allen ReacTV	Feb 26, 2017	2,700	118	MARCOS GOLD: SEKRETO SA LIKOD NG 192,000 TONS OF GOLD (Marcos Gold: The secret behind 192,000 tons of gold)	Aug 12, 2019	147,000
Toto Bee	Feb 22, 2015	15,500	123	BAKIT PINAALIS SI MARCOS mini documentary (Why was Marcos driven out of power mini documentary)	Feb 27, 2018	847,000
JDBros	March 12, 2015	147,000	70	MARCOS GOLD CAN SAVE THE WORLD 987 BILLION DOLLARS AND MILLION TONS OF GOLD IN THE PHILIPPINES	Aug 31, 2017	2.005M

topics in social media by using them as tags. This is possibly to increase their visibility in YouTube's homepage or suggested videos (e.g., Coronavirus, other political issues such as the non-renewal of the broadcast franchise of the country's largest network, ABS-CBN, and other global events). Nonetheless, the clear focus of these channels, with a dedicated number of videos, is to promote the name and legacy of the late Philippine dictator Ferdinand Marcos, while touching on general historical events. This is achieved by advancing myths and conspiracies about Marcos and his presidency (outright disinformation) and strategically framing historical facts and accounts to their benefit (distortion of reality). Buttressing this narrative is the continuity of Marcos' legacy through the agenda of the Duterte administration and through the election of his son Bongbong Marcos for President. The brokers skilfully promote this political agenda through their videos and channels, while monetising directly from YouTube.

Brokering Politics, Broadcasting Historical Revisionism on YouTube

Aspirations: Past, present and future

Political brokerage anchors on political and national aspirations of Filipinos, and which brokers capitalise on – appropriating a blending of fact and fiction in their discursive styles. As argued by Curato (2020, p. 77), "historical revisionism through social media" is rooted in "'deep stories' of ordinary Filipinos about how they view themselves, their personal circumstances, and their relationship with the nation." We zoomed in to the aspirational narratives being sold by these political brokers to understand how they penetrate the Filipinos' psyche through stories that will connect to their desires and imagination for the nation. The aspirational tropes that emerged from our study bring in narratives of "what was" and "what could be" – a sense of national pride ("we were once a great nation"), wishes for a "great leader," regional competitiveness and independence from previous colonial controls ("we can be an independent nation, free from bullish American influence"), and a "true and working democracy" ("we need a democracy that will work for the Filipino and pave the way for social justice even when it might stifle political liberties"). It can be noticed that these tropes work at the level of ideology, and at a practical level, experience.

1.   National pride ("We have always been a great nation")
The first key aspirational trope is the perpetuation of the narrative that
the Philippines was a great nation during the time of former President
Marcos. Notably resonant with former U.S. President Trump's tagline, "Make
America Great Again," this draws on a pre-colonial myth of the Philippines
as a *"Kingdom of Maharlika,"* a noble nation, that was enriched and made
continually vibrant by Marcos with an unparalleled level of social and
political development that was apparently reversed when Marcos was ousted
from power. This was articulated by brokers as they emphasised Marcos'
belief in the greatness of the Filipino, "As a former Philippine President, he
would always say, 'We shall make this nation great again.' Why? Because
he has always believed in the greatness of the Filipino" (Sangkay Janjan
TV, 2019). Marcos' "book-length" (Jevera PH, 2019) economic and cultural
accomplishments would then be displayed – from bridges to power plants
to cultural centres and to universities, which, according to brokers, "remain
to be of important value 'til the present" as a marker of possibility if the
right leaders are elected (SangkayJanJan TV, 2019). These achievements
would then be juxtaposed with squalor, poverty, and failed democracy
experienced during the succeeding administrations, particularly that of
Marcos' long-time political rival, Corazon Aquino (Sangkay Janjan TV, 2019).
Interestingly, the brokers would see Rodrigo Duterte's rule as a good revival
and continuation of Marcos' legacy (Birador HQ, 2020).

2.   A decisive, benevolent, high-achieving President ("not just as the greatest
      leader of the Philippines, but of the entire world")
The second trope of videos involves the promotion of former President Marcos
as a benevolent and exceptional President, unparalleled in his accomplish-
ment as the country's leader. Marcos is identified, "not just as the greatest
leader of the Philippines, but of the entire world" (Sangkay Janjan TV, 2019;
also in Jevara PH, 2019), and whom other global leaders revered. Marcos
has laid out the landscape for this development before it was halted: they
would emphasise the development of the Bataan Nuclear Power Plant built
by Marcos to sustain the country's energy needs, a project that has been
considered to be a risky investment and significant wastage of government
funds, but to the brokers is considered the marker of Marcos' economic
foresight and intelligent governance (OHJAYCEE, 2020; Filipino Future, 2020)
    To promote Ferdinand Marcos' image as a benevolent leader importantly
requires challenging the controversies, issues, and criticisms directed at him
and his regime. To this end, YouTubers would strategically use the videos
to be in defence of Marcos by either (1) positioning him as a "victim" of

black propaganda instigated by opposition leaders, elite establishment and foreign powers (e.g., EDSA People Power as a front an orchestrated ouster by the elite opposition and the US) or by (2) justifying his policies and actions as logical and/or inevitable given the circumstances at that time (e.g. rise of communism as a justification to Martial Law). For example, in a video, *"Were People Really Abused During Martial Law?,"* YouTuber Sangkay JanJan (2019, March 21) counters official gazette records of human rights violations and torture. He argues that the news and supposedly historical records pointing to torture and atrocities during Marcos' Martial rule were not true. He justifies, based on his "own research," that "only the communists who were bringing down Marcos' good administration through violent means" were punished for civil disobedience. He emphasises that Filipinos should be careful about these fabrications that demonise Marcos, narratives that were crafted by political opponents angry at Marcos and seeking to taint his good reputation for their own selfish gains. Claiming to be people's "ally of truth," Sangkay JanJan invites his viewers to subscribe to his videos where they ought to get a regular offering of well-researched historical facts.

3.   "A true, working democracy"
The third aspirational trope is the vision of a "true, working democracy," a "democracy underscored by discipline, equality and respect" (Sangkay Janjan TV, 2019). It is a vision of democracy "where the Philippine government is organised and with functioning laws that are obeyed by a disciplined citizenry" (Jevara PH, 2019). The brokers would explain that contrary to critics' accusation that Marcos is anti-democracy, his brand of democracy is the one that was most just, and "if Marcos' plans were only actualised, the Filipino people would have benefitted from it immensely" (Jevara PH, 2019). This democratic vision of an egalitarian society would then be juxtaposed with images of Marcos' projects across various development areas (Jevara PH, 2019; Sangkay Janjan TV, 2019) (See Figure 6.1). This would include "infrastructure development that benefits the people," "giving away of lands to farmers for sustainable agriculture" (Filipino Future, 2020), and the "sequestration of the riches from oligarchs" (such as the Aquinos) for redistribution to the people (Filipino Future, 2020; also in Toto Bee, 2018).

4.   Regional competitiveness and independence from the United States
"We were once economic superpowers"; "We can be an independent nation once again"
    Aligned with the enumeration of Marcos' achievements, pro-Marcos brokers painted a picture of nostalgia over the glory days during Marcos'

leadership that positioned the nation as an "economic superpower" equal, if not more vibrant, than the United States or Singapore. Looking back at this version of history, the brokers would then lament this lost glorious past, while emphasising a vision of "what could have been," if Marcos was allowed to continually rule: "the country would be a greater superpower now than America and even in comparison to our neighbours in Southeast Asia" and "we will no longer rely on the dollar" (Filipino Future, 2020). What is interesting in this particular aspirational narrative was the emphasis on political and economic independence from previous colonial powers (i.e., United States). The reference to the U.S. is therefore meaningful as it pertains to an internal dialectical tension of seeing the coloniser as a symbol of progress while at the same time wanting to overcome and supersede it. This trope of economic greatness and independence was also applied by the brokers to issues that Filipinos can easily relate to in their everyday lives: that of overseas labour migration. Connected to the vision of an economically independent nation is going beyond "the need to go overseas to become a slave in other countries" (Sangkay Janjan TV, 2019) and which brokers purport to be easily realisable if the Marcos' would be allowed to continually rule.

An interesting myth related to this aspiration of economic independence is the "Marcos gold." The narrative begins from a pre-colonial myth of the *"Kingdom of Maharlika,"* pertaining to a Filipino family that amassed gold in the Southeast Asian region, and which was later on apparently inherited by Marcos ("Marcos' gold") (Filipino Future, 2020; AllenReacTV, 2019; JDBros, 2017). The argument by the brokers is that the tons of supposedly inherited gold, which was deposited by Marcos in Swiss banks, was actually meant to earn and eventually taken back by Marcos for distribution to Filipinos and for pumping the Philippine economy towards vibrancy. Unfortunately, Marcos' "noble plan" was disrupted when he was ousted by the Aquinos – but the brokers would emphasise their vision of "what could be" if Marcos had been given the chance to push through with his plan, where the country becomes "economically independent even without foreign investment" and where other so-called "superpowers will lose their power and get overshadowed by the Philippines" (Filipino Future, 2020).

This story, told across several videos, would wrap up nicely with an endorsement of Marcos' son, Ferdinand ("Bongbong") Marcos, Jr. for the Presidency. Attacking the opposition, a vlogger further quips, "they are the only ones who do not want Bongbong Marcos, but the majority of the Filipino people are very excited to have the Marcos' rule the country once again as President, so that we can revive our country's reputation as a noble land of the Maharlika" (Filipino Future, 2020). Brokers would argue that reinstalling another Marcos would be the only way the wrongdoing in relation to the noble Marcos gold could be retrieved and used for the country's national development aspirations (JDBros, 2017). This further reveals the YouTuber's brokerage role. It tells us that the YouTube videos are not mere musings of one regular YouTuber's research on history. Instead, the brokers and their videos can be seen as advancing political propaganda while bridging a politician with viewers – as voters. Their potency lies further in the corroboration and amplification of the message across the videos and channels of different brokers.

Discursive styles

In *The Place of Media Power*, Couldry (2000) analysed the legitimation of media through five interlinked processes: framing, ordering, naming, spacing and imagining. Pro-Marcos brokers engaged these discursive styles to advance political and national aspirations, incorporated with semantic and logical tactics attuned to the platform vernacular. These discursive styles were intricately intertwined in the videos, using a combination of narrative, visual and audio techniques to create a palatable and persuasive political rhetoric for Filipino YouTube audiences.

The shifting temporalities of the political aspirations necessitated the brokers to create imaginaries and representations of the glorious past and future of the Philippines under a Marcos government. *Imagining* techniques were used to aid audiences, particularly young viewers, to visualise what life was like during the Marcos regime. Brokers characterise the era in tangible and familiar ways by presenting the images of major infrastructures like trains, bridges and powerplants, staging photos of Marcos with prominent world leaders, and reciting the affordable prices of basic goods at that time. The same tactic was used to project an alternate reality if Marcos stayed in power, where "there would be no informal settlers... no dirty roads... no one would leave to work abroad" (Sangkay Janjan TV, 2019). These imaginaries were accompanied by images of vectorised urban landscapes and city centres of neighbouring ASEAN countries to show what the Philippines would look like as a developed nation. They used aspirational images of progress and development in other countries (i.e., Singapore) which is familiar to geographically mobile and Internet savvy Filipinos.

Intertextuality and mythmaking were also used as imagining tactics to weave the brokers' narratives into the present cultural milieu. The brokers consistently emphasise Marcos' political platform of "Make this nation great again" to mirror Donald Trump's campaign slogan of "Make America Great Again." The veneration of Marcos' "last words" (OHJAYCEE, 2020) were akin to Jesus Christ and other revered personalities. Several myths were perpetuated about the alleged billions of gold bars owned by the Marcos family to explain accusations of their ill-gotten wealth (Allen ReacTV, 2019), and of Marcos' last will to distribute the wealth to Filipinos (Filipino Future, 2020). Conspiracies were also used to undermine the legitimacy of the EDSA People Power by propagating that it was an orchestrated plot of the elite establishment (Toto Bee, 2018; Filipino Future, 2020).

Through the use of *framing* devices, the brokers were able to reconstruct Marcos' legacy as a pretext to nurture the aspirations for a progressive nation. One prominent tactic was emphasising historical details that elevate Marcos' accomplishments while silencing their underlying issues and consequences. In the litany of Marcos' projects, the Bataan Nuclear Power Plant was characterised to be the "answer to the rising demand for electricity" and to "minimise the country's dependence on imported oil" (OHJAYCEE, 2020), but surrounding issues concerning corrupt financing schemes and potentially hazardous implications that led to its closure were not mentioned. Marcos supposed "win" against democracy icon Corazon Aquino in the 1986 snap election was portrayed as legitimate and devoid of the issues of rampant electoral fraud (Tinig PH, 2020; OHJAYCEE, 2020).

Detractors of Marcos and supporters of the opposition party were consistently framed as the ignorant "others" who impeded the actualisation of Marcos' legacy. Brokers articulated that the anti-Marcos refuse to see the merit of Marcos' projects and they "let themselves be fooled" (Sangkay Janjan TV, 2019) by the opposition's propaganda. Despite historical records documenting abuses during the Martial Law regime, the brokers berated the victims of human rights violations for "baselessly" condemning Marcos, aligning them with the critics of the government, "Are the human rights abuse victims and idiotic Aquino followers really after justice or they are just finding someone to blame?" (Toto Bee, 2018). After showing an interview with an elderly person who attests to the Martial Law regime being a highly peaceful period, another broker further nullified the calls for justice of the claimants: "As you have seen, no one was really abused during the Martial Law. This was just made up by Marcos' detractors to taint the great leader's legacy!" (SangkayJanJan TV, 2019). The brokers also engaged in normalising the claim that the protesters during EDSA People Power revolution were "minority" and that "majority" of the Filipinos are supportive of the Marcoses. Dissenters were framed as "a few people angry at the Marcos government" (Tinig PH, 2020). Further, brokers claimed that this minority are "the only ones who oppose Bongbong [Marcos], but the majority of the Filipinos are excited to vote for a Marcos again in Malacañang [Presidential Palace]" (Filipino Future, 2020).

The brokers' framing strategies portrayed Marcos as the best leader the country ever had, but who was the "victim" of decades-long propaganda set-up by the opposition party, members of the elite and foreign powers. Brokers insisted that the allegations against Marcos being a dictator, a mass murderer and a kleptomaniac were instigated by those threatened by his plans to dismantle the oligarchy (Sangkay Janjan TV, 2019). Marcos allegedly had to endure "the intense humiliation and insults, and the vile treatment of him and his family, because he was thinking of the future of his beloved land" (Filipino Future, 2020).

The brokers also used *ordering* and *spacing* strategies to define the ideal nation using strategic parallelism, oppositions or juxtapositions. They used contrast to create an "order" of importance of achievements that work to elevate Marcos' legacy, while creating a spacing between these grand achievements and the failure of other administrations that succeeded Marcos. There was an abundant use of lists to display the grandness of infrastructure projects during the Marcos era, with one broker spending ten minutes of his video dedicated to enumerating the achievements of the late dictator (Jevera PH, 2019), of course with affective musical accompaniment. They also indirectly offered their characterisations of best and poor governance. The economy and society during Marcos' time as President was purportedly "the best and the finest [time] of the country compared to the pre-Marcos and post-Marcos era" (Toto Bee, 2018). The brokers particularly claimed that Corazon Aquino's rule diminished the growth that the Philippines experienced under Marcos (Toto Bee, 2018; Sangkay Janjan TV, 2019; OHJAYCEE, 2020) while her son and former President, Benigno Aquino III, has failed with his reforms that the Duterte administration was able to redeem (Birador HQ, 2020). Another analogy that the brokers mobilised to support this comparison was equating the opposition with the oligarchs, and Marcos and his allies with the nation's interests. The Aquinos, being part of the elite family, allegedly restored the power of the oligarchy by giving them back ownership of major companies that Marcos sequestered during Martial Law (Toto Bee, 2018).

Through both narrative and visual devices, the brokers delineated the Philippines' state of destitution since Marcos died with the economic progress of our Southeast Asian neighbours. Narrated through images of progress, brokers claimed that "the countries that Marcos helped in the past, like Vietnam, Thailand, Indonesia, Korea and Singapore are now thriving nations. Meanwhile, the Philippines, who were mere beneficiaries of Marcos' great governance, have apparently superseded the Philippines as the country further descended into poverty and never recovered" (OHJAYCEE, 2020).

The language, expression and tone of the videos revealed the semantic styles of the brokers in producing affectively charged propaganda. Through naming and labelling, Marcos was persistently portrayed in hyperbolic ways. Brokers professed that Marcos was the "greatest leader in the whole world)," while his achievements "can fill up a whole book." One broker took a different approach and has conceded that Marcos was an authoritarian and a dictator but redefined the term as someone with the "authority or capability to be a President... [who] uses to hold power and manage the country effectively" (Toto Bee, 2018).

The semantic choices of the brokers characterised the Marcoses as having the moral ascendancy over their political opponents. On the one hand, their word choices positioned former President Cory Aquino as unscrupulous and vindictive, complemented by images of her projected like a villain surrounded by her "oligarch" allies. In reference to her proclamation as President, brokers asserted that she and her party have "stolen the government from Marcos" (Tinig PH, 2020). Marcos' projects were also allegedly "sold by the next president [Aquino]" (Jevara PH, 2019). The vilification of Aquino was used to elevate Marcos as the magnanimous leader and brokers frame his ousting as a "lesson" to Filipinos to "stay loyal" and "not hastily abandon decent leaders" (Sangkay Janjan TV, 2019). There was also an excessive use of crass and derogatory language against Marcos' critics in these brokers' videos. They engaged in name calling, such as idiot, fools, simpleton, among others, and condescend them as ignorant of history, politics and economics. Brokers chastised Marcos' critics as "stupid and blind for believing in them [opposition] and allowed for these 'animals' to successfully enact their plan" (Birador HQ, 2020). They also ridiculed and attempted to discredit people and events that led to Marcos' removal from power such as the People Power Revolution of 1986 (Toto Bee, 2018).

On the other hand, they used unique labels to address their loyal viewers and subscribers. They used local language that pertain to belonging, solidarity, and a sense of loyalty to brotherhood or friendship (i.e., *mga ka-sangkay, mga kabayan, mga birador, ating channel, solid Sangkay*). Notable was how the same semantic strategies mimicked the style used by major broadcast networks in the country to address their loyal viewers (i.e., *mga Kapuso* or beloved or "heart-mate," "*mga kapamilya*" or family, "*kapatid*" or brother).

Embedded in the discursive styles above were logical fallacies used in making claims about Marcos' legacy and the failure of other administrations. The brokers discreetly "sandwiched" falsehoods in between historical facts and events. They followed the sequence of introducing basic textbook information about Marcos, followed by factual distortions and opinion, then bundling these together with indisputable evidence. Brokers methodologically presented Marcos' self-evident infrastructure projects and then transition to discussing their claims about the economic conditions of the country under the Marcos regime (Jevara PH, 2019). Regional events such as the rise of communism in Southeast Asia were also used by the brokers to justify the proclamation of Martial Law in the Philippines (Toto Bee, 2018). The brokers also engaged in false attribution to build a rosy picture of the Philippine economy during the Marcos regime, and economic demise after he was ousted. They asserted that "after Marcos' death, the country went into shambles," indicated by economic decline, absence of meaningful reform, and environmental degradation (OHJAYCEE, 2020; Sangkay Janjan TV, 2019), without recognising that much of the country's sorry economic state was inherited from Marcos' leadership that spun for over several decades. These claims were a set-up for the brokers to build the appetite for the Philippines to be the country it once was by "choosing to re-install the Marcoses in Malacañang through their sacred vote" (Filipino Future, 2020). The unique ways in which the videos blend fact and fiction, commentary and facts, speculation and direct manipulation, and information and jokes, create a composite narrative that is challenging to fact check or dispute.

Credibility-building strategies

Pro-Marcos brokers on YouTube wielded their credibility not solely on "expertise" or privileged relations (Stovel & Shaw, 2012), but on the accessibility and relatability of their political rhetoric. They achieved this by privileging the subjective construction of knowledge through their position as "ordinary" citizens conducting sincere research on history, but in a manner that imitates the devices of institutional sources of knowledge. While they leverage on the climate of heightened distrust of intellectual authority, they also built a distinct brand of affective political brokerage compelling to parties with vested political interests and a disaffected public.

The brokers expressed their rejection of conventionally accepted history established by traditional gatekeepers of knowledge such as academia and media. Through their videos, they claimed to "divulge the true history of the country that history books or universities forcibly concealed" (*isisiwalat ko ang mga tunay na kasaysayan ng ating bansa na hindi masabi ng mga libro ng kasaysayan o sa mga unibersidad na pilit tinatakpan ang katotohanang ito*) (Sangkay Janjan TV, 2019). Instead, they endorsed alternative "insider" knowledge that history books purportedly covered-up to protect particular interests. There was a repeated invocation of how the brokers' version of history is "unknown or hidden from many" (*lingid sa kaalaman ng marami*) (Filipino Future, 2020). In one video, the broker even leaned towards the camera and whispers, as if sharing a secret or gossiping. Interestingly, this renunciation of anything "official" is not encompassing: the brokers still used academic and media reports as references, but they cherry picked or obfuscated the details they pick out from them.

By shifting the onus of judging the integrity of information from institutions to the individuals, brokers derived their reliability from their subjective curation of knowledge. Brokers proclaimed that "they have found" (*natagpuan ko*) (Sangkay Janjan TV, 2019) their evidence from executing their own research and imply personal accountability in the validity of their claims. Crucially, they promoted an "illusion of discovery" – that their historical knowledge "emerged" from research and seek to cascade the same illusion to their viewers. This heavily relied on the brokers earning their viewers' trust, not from demonstrating the soundness of their investigation, but from creating relatability through affective communication. The brokers introduced themselves at the beginning of their videos and emphasise that affinity to their viewers as "fellow millennials" (*mga kapwa kong millennial*) (Sangkay Janjan TV, 2019) or "fellow citizens" (*mga kababayan*) (OHJAYCEE, 2020). They also engaged in strategic concessions, to present themselves as discerning, if not neutral. More importantly, the brokers built their relationship with the viewers by "recognising their right to discern what and who was President Marcos" (Tinig PH, 2020). By privileging a do-it-yourself approach to knowledge building, the brokers were able to deflect apprehensions and interrogations about the claims being supported by documented evidence and records.

Despite the brokers' repudiation of traditional knowledge sources, they drew from the same playbook in packaging their brand of brokerage. Working within the conventions of the educational video genre, the brokers always began with established facts about Marcos and details of historical events. They also presened references and evidence in their narration, regardless of their quality or validity. Their narrations were always delivered in a confident, as-a-matter-of-fact tone. Attached to their brand and in their videos are images and signs that were generally associated with knowledge and information, such as brains, a formula and computers (See Figure 6.1). Further, the brokers attempted to present themselves as professionals by creating opening and closing billboards to elevate their content from the rest of the amateur videos and into their own media brand.

Figure 6.1   Still from a YouTube video (JevaraPH, 2019). Credibility-building strategy: Broker uses imagery related to knowledge and information in the videos' logo and branding.

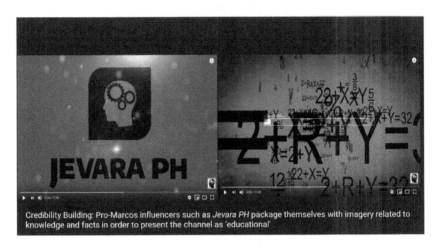

Credibility Building: Pro-Marcos influencers such as *Jevara PH* package themselves with imagery related to knowledge and facts in order to present the channel as 'educational'

Platform-specific engagement tactics

The brokers used engagement tactics specific to the affordances of YouTube. For example, all brokers (except Toto Bee) always started and ended with a call to subscribe to their channel and press the "bell button" to receive notifications from them. They would also actively call to comment, like and share their videos/channels with the possibility that viewers would be featured or recognised through "shoutouts" in the brokers' videos, which appears to be projected as an aspirational target for some of the followers. One particular broker, Birador HQ, explicitly targeted a specific number as he mobilised viewers to like his videos: "Let's work together to reach 500 likes for this video!" They would also recommend viewers to watch other videos in their channel (Sangkay Janjan TV, OHJAYCEE, Jevara PH, Toto Bee). Loyalty appeared to be a prominent feature for the brokers' engagement tactics, affectionately calling on the names of some "loyal" or "solid subscribers" and wishing them, "long live!" (*mabuhay!*). As a common strategy among micro-celebrities, community building included soliciting topics for next videos in the comments section and selective revelation of their private lives through the sharing of selfies, family photos or footages, and travels and injecting them in their political narratives.

Behind the YouTube interface were keyword tags attached to each video. These tags were only viewable through the page source, which contained the code for the web page. Notably, brokers used keyword tags that were semantically close ("facts," "myths," "*kwento*" or story) and discursively related to tap into parallel searches using these keywords. The brokers used the tag function generously to expand the topics their videos can be associated with. Apart from the keywords related to the search term, Sangkay Janjan TV also included names of prominent political personalities (Ninoy Aquino, Corazon Aquino, Rodrigo Duterte) and tagged local YouTubers (Jamill, Cong TV, Wil Dasovich) with millions of subscribers, potentially to ride on their popularity. They also appeared to be strategic in their titles. For instance, they highlighted the existence of concealed or undisclosed historical accounts through the use of expressions such as "Untold History" and "Hidden Truth." These phrases were common click-bait headline tactics in social media, but it could also work to interpellate viewers into subjects who are receptive to the historical revisionism and the systemic undermining of traditional gatekeepers of knowledge, more broadly. Notably, our data showed that brokers are well attuned to the expectations for micro-influencers in this social media platform, and this was manifested by the attention to posting regular videos in their channel and creating videos on key trending contemporary topics that are not necessarily closely politically oriented.

YouTube as Political Broker

Amid the tactics specifically engaged by the brokers, YouTube, on its own, plays a crucial role in making visible, ordering, and curating the videos for a user/audience (Soriano & Gaw, 2021). First, the visibility of a video was highly contingent on it appearing on the search result. In each iteration, the search results produced a slightly different set of videos (some new videos appear, some disappear) in slightly different rankings (a top five video may be pushed down the list outside the top 10 in the next search, e.g., Jevara PH's video was shifted from four to nowhere in the top 10 in the next search). However, it is important to emphasise that from the keywords "Marcos History," there was only one consistently appearing video that is professionally produced (ML Chronicles documentary) and the rest of the top 10 were all amateur videos of political brokers that lean towards a pro-Marcos stance.

Our data showed that YouTube's video recommendation system narrowed down the selection of videos into content produced by amateur content creators and broadens it to include videos with the same fringe themes (more conspiracy videos, folklore, hoaxes, etc.). After multiple iterations, the same pro-Marcos propaganda videos that appeared in the search results and in other recommendation lists were recommended repeatedly, implying a networked connection between them (either through the users' viewing history, the video's content, and the general audience behaviour). YouTube also offered affordances that help the broker gain visibility and attract audiences in particular ways. For example, once a viewer clicks on the "subscribe" button, new videos uploaded by the brokers appear in their subscription feed. However, only when the user clicks on the "bell" button will they get a notification for every new video uploaded by the brokers. This is why the brokers in this research always have two "calls to action" – subscribe and click on the bell button to be able to directly re-engage their viewers.

On top of the views and the watch time of the videos, audience engagement factored in how YouTube computes for the ranking of videos. This means that even without massive viewership, a video can be scored higher by the algorithm because of its engagement metrics. The brokers we analysed also always encouraged their viewers to like, comment and share their videos, possibly to push their videos up the YouTube ranking, and consequently obtain greater monetisation. Further, although the research did not focus on the comment section, we did observe that it is part of the process of political brokerage, although this certainly can be the subject for more detailed investigations in the future. The comments in the videos appeared to express strong pro-Marcos sentiment, directly affirming or extending the content/claims of the video. The comments strengthened the credibility of the video through affirmation, and the commenters potentially served as another broker supporting the dominant discourse promoted in the videos.

YouTube and Political Brokerage

This chapter examined YouTube as facilitating the construction of political and historical discourses by an emerging set of political actors whom we call brokers. First, we highlighted the importance of connecting to the Filipino aspiration and YouTubers mobilise "mythmaking" in the creation of narratives that pander to people's multiple aspirations and imaginaries for the nation.

Although we examined the production of myths in relation to Marcos' history on YouTube, this process is not new (McKay, 2020, Reyes & Ariate, 2020; Reyes, 2018). Previous works have pointed out that discourse on Ferdinand Marcos is often punctuated with such mythic reduction (Reyes & Jose, 2012/2013, p. 4). For decades, Marcos has defied a sober summation, with the pro-Marcos camps regurgitating the myths peddled by Marcos' propagandists and manufacturing their own glowing constructs of the "Greatest President Ever," and the anti-Marcos camps struggling to counter these myths with unadorned statements of fact from historical research (Reyes, 2018; Reyes & Ariate, 2020).

The potent combination of stories that connect to the Filipino's aspirations combined with a mastery of the platform vernacular through the construction of "ordinary expertise" and "self-discovery," make untruths and half-truths palatable. Especially when told by a network of commentators, these obtain a semblance of reality. YouTubers engage the myths peddled by Marcos propagandists through self-published books, but make these more visible, relatable, and affective through discursive styles and engagement tactics possible given the affordances of YouTube.

Importantly, beyond the production of content, we emphasise that YouTube facilitates a brokerage process where content creators can connect dispersed information – factual or not – that ultimately allows them to also intermediate between political actors and citizens. YouTubers are intermediaries between users/voters/publics and actors with political agenda who can benefit, directly or indirectly, from the brokerage process. We may see them as plain content creators, but they essentially create a bridge between two poles, making propaganda and political campaigns appear like sincere knowledge pursuits in this DIY cultural media economy. It is by appearing to possess "in-group identification" and therefore seemingly "biased" (Stovel & Shaw, 2012) toward ordinary citizens by presenting themselves as "ordinary experts" or "political and historical commentators from the masses" who understand the Filipinos' deep aspirations – that their brokerage role assumes believability and legitimacy. This appears to be more potent in the context of the Philippines, which is characterised by constant political turbulence, a weak state, declining media trust, and a public that is simultaneously entertainment and politically savvy (A. Pertierra, 2021).

What appears to be the convergence point that can be drawn from the findings is how brokers anchor on postcolonial political and national aspirations of Filipinos. The aspirational tropes that emerge from our study bring in narratives of "what was" and "what could be" if Marcos was not deposed from power or if the Filipino people elected another Marcos to the Presidency – a sense of national pride, regional competitiveness, an economically robust nation, or independence from previous "colonial controls" – which their political narratives anchored on. As Benedicto (2013, p. 26), in describing one of the key markers of Marcos' grand infrastructures, the Manila Film Centre, argued, Marcos' infrastructure projects

...engendered a sense of both nostalgia and futurism by positioning itself as a force that could, at once, transport the city back to the glories of a protonational past and propel it toward advanced states of economic development. Moreover, the structure's departure from and adherence to the tenets of architectural modernism mirror the precarious position occupied by the Marcos regime as a postcolonial dictatorship. Its architecture abides by central "international" principles such as the rejection of adornment and frivolity, but it also affects spectacular excess through scale, height, and the starkness of its contrast with the "thirdworldness" of metropolitan Manila.

As shown in this chapter, the mythical narratives of the brokers in reconstructing Marcos' glory days while sanitising accounts of torture and human rights violations appear like its other infrastructure projects such as the Film Centre, "a built environment that bears the signature of lost time" (Benedicto, 2013, p. 28) and that elicits a resurrection for those who subscribe to the dictatorship's promise of modernity. Yet, the narratives also exemplify imaginaries of progress, optimism, and globalism that attempt to erase feelings of terror and dread during Martial Law. These contribute to ongoing efforts of erasing the turbulent legacies of the Marcos regime and especially, of its impact on the victims of abuse during Martial Law that is well documented in official historical accounts. Importantly, it also carves out pathways to the return of the Marcos' to the Presidency.

Secondly, political brokerage illustrates the ambiguity of political influence and historical knowledge construction in the context of YouTube. Political influence is constructed through the legitimation of particular political narratives as knowledge and attempts to erase a turbulent period in history for a political agenda. Anchoring on our point about crystallising aspirational narratives, brokers rarely present a simple blatant manipulation of knowingly false information. The more common strategy is the creative blending of fact and fiction, conducted through the use of some historical facts and inserting these with a slew of the commentator's interpretations and opinion, of creative discursive styles, or using semantic strategies to blur facts to make narratives compelling. The YouTube videos we analysed would run for three to 10 minutes, giving the broker ample time to expound on arguments and pepper these with video footage, real images of people and events, and rousing music that create a sense of believability. Brokers are aware that people would not be fully convinced with single posts on Twitter or Facebook – they need a composite of stories of despair, frustration, hope, and aspiration well captured in YouTube videos that could trigger affective response. Accompanying these audio-visual narrative styles, we point out that political brokerage is embedded in the techno-sociality of communication, highlighting the interactions between the platform and its users that makes this possible.

We now turn to the platform YouTube which configures the action of brokers through its composite features and affordances, allowing the brokers to advance a political agenda through historical revisionism. The platform serves as a space for the flourishing of "proxy makers of authority" or ordinary people broadcasting content with a political agenda (Soriano & Gaw, 2021). The shareability of content affords its distribution and consumption across different platforms. Notably, by feeding content to viewers, YouTube advances and normalises the influencer's version of Philippine history. Importantly, the financial incentivisation of influencers sustains a marketised and politically charged environment, and this is reinforced through loose platform governance. Platforms are political because they are in the business of governing people, objects and discourses within the boundaries of their infrastructures (Bucher, 2018). The product of such governance is its potentiality to construct new realities, and political brokers take advantage of this by creating a network that builds, propagates, and cements their political narratives without being subjected to the same scrutiny of traditional gatekeepers. In other words, the brand of political brokerage that works through deception and fabrication might not be as easily permissible in mainstream forms of media, but they thrive in platforms with porous governance structures such as YouTube.

Paying ethnographic attention to the working of the political brokerage process on YouTube illuminates the broader structure that constructs political discourse in the contemporary digital environment. Brokers are made visible, their content categorised, and are allowed to be subscribed to – imbued with their personal (or politically motivated) interpretations of history and political vision – and as a result of this socio-technical interaction they are able to connect to viewers whose aspirations align with this vision, or whose political views can be shaped and fortified through sustained exposure to such content. The platform YouTube, through its algorithmic recommendation system and other affordances, can therefore also be seen as a "broker," functioning as a "middleman" (Lindquist, Xiang & Yeoh, 2012; Stovel & Shaw, 2012) that bridges the Filipino YouTube public to these brokers. This double layering of brokerage has important implications for the role of social media in examining political engagement.

Participatory culture has made it possible for content and information sources to be seen and shared outside the confines of regulation and vetting of the media and traditional forms of expertise and this allows for changing notions of trust and expertise to emerge. This works well against the broader narrative of casting doubt on the credibility of media and traditional expertise. When the same narratives are reinforced by brokers and this becomes consistent, persistent, and corroborated by several brokers and across platforms, audiences may begin to trust their narratives. The brokers' ordinariness and relatability makes narratives more believable and frees it from a feel of propaganda. What is also apparent is that YouTube itself as a platform is expertise agnostic, as YouTube's algorithmic recommendation system gives visibility to brokers regardless of the quality of the evidence that they present. When one clicks on a broker, the platform continually embeds the user to more recommended videos of the same or other brokers (see also Soriano & Gaw, 2021). On Facebook or Twitter, content is anchored on social relationships, as videos and content are shared within a person's sphere of trust. On YouTube, a sense of trust can be established through the careful balancing act of manufacturing expertise and assumption of ordinariness. This implies a logic of credibility and trust building different from other social media platforms.

We now end with some points for further thinking about the implications of political brokerage on YouTube. Certainly, YouTube plays a large part in these micro-influencers' prominence sans the scrutiny of traditional gatekeepers. Anyone can create a channel and produce historical claims that are not restricted by the limitations of bandwidth, airtime, and editing or regulation by historical institutions. The platform made it possible for information to be seen and shared outside the confines of traditional expertise and regulation, but it is also being appropriated by forces to advance politicised critique and partisan political commentaries. We ask, How do we treat these "historical" narratives and political brokerage on YouTube? Are they to be considered as entertainment? As knowledge? As disinformation? The very ambiguity of the nature of content on YouTube and how we ought to deal with them allows them (as well as the platform) to elide responsibility. Further, despite common calls for media literacy, this hybrid media environment makes it more and more complex for audiences to make sense of what counts as historical knowledge versus political propaganda. Looking more closely at how audiences respond and engage with these influencers would be an interesting research to further explore. Finally, if we see these influence strategies as effective, perhaps historians and historical institutions can collaborate with artists and content creators in using YouTube's affordances and the same compelling strategies to advance evidence-based historical material.

Political brokerage brings together and organises disparate actors and political content amidst an overabundance of content on social media and the crevices in existing Philippine political infrastructures. The scope of the study does not allow us to affirm whether these YouTube brokers are part of disinformation infrastructures (although some evidence may signal an orchestrated strategy) or whether they are motivated by personal ideological leanings. Nonetheless, we argue that whether part of a disinformation architecture (J. Ong & Cabañes, 2018) or not, they are able to powerfully broker a political agenda that benefits them and particular political interests. The brokerage process attends to gaps in information by visibilising preferred political narratives that are curated for a user's needs and personal political tastes, thereby allowing YouTubers to become political intermediaries. In another, this political brokerage process breeds legitimisation of content into knowledge without accountability, while allowing the brokers to gain economically from the process.

References

Abidin, C. (2017). #familygoals: Family influencers, calibrated amateurism, and justifying young digital labour. *Social Media + Society, 3*, 1–15. doi:10.1177/2056305117707191

Aguilar, F.V., Jr. (2019). Political conjuncture and scholarly disjunctures: Reflections on studies of the Philippine state under Marcos. *Philippine Studies: Historical and Ethnographic Viewpoints, 67* (1), 3–30. doi:10.1353/phs.2019.0004

Airoldi, M., Beraldo, D., & Gandini, A. (2016). Follow the algorithm: An exploratory investigation of music on YouTube. *Poetics, 57*, 1–13. doi:10.1016/j.poetic.2016.05.001

Arthurs, J., Drakopoulou, S., & Gandini, A. (2018). Researching YouTube. *Convergence: The International Journal of Research into New Media Technologies, 24* (1), 3–15. doi:10.1177/1354856517737222

Aspinall, E., Davidson, M., Hicken, A., & Weiss, M. (2016). Local machines and vote brokerage in the Philippines. *Contemporary Southeast Asia, 38* (2), 191–196.

Benevenuto, F.; Duarte, F.; Rodrigues, T.; Almeida, V. A.; Almeida, J. M.; and Ross, K. W. (2008). *Understanding video interactions in YouTube.* In Proceedings of the 16th ACM International Conference on Multimedia, MM '08, 761–764. New York, USA: ACM.

Baustista, V.F. (2018). The Pervert's guide to historical revisionism: Traversing the Marcos Fantasy. *Philippine Studies: Historical and Ethnographic Viewpoints, 66* (3), 273–300. doi:10.1353/phs.2018.0026

Bucher, T. (2016). Neither black nor box: Ways of knowing algorithms. In S. Kubitschko & A. Kaun (Eds.), *Innovative Methods in Media and Communication Research* (pp. 81–98). Springer International Publishing. doi:10.1007/978-3-319-40700-5_5

Burgess, J., & Green, J. (2018). *YouTube: Online video and participatory culture* (2nd ed.). Polity Press.

Chang, H.H., & Ahn, S.Y. (2020, May 23). #YouTubeAndrewYang: How a random man became a presidential candidate [Paper presentation]. International Communication Association Virtual Conference.

Couldry, N. (2000). *The place of media power: Pilgrims and witnesses of the media age.* Routledge.

Curato, N. (2019). *Democracy in a time of misery: From spectacular tragedies to deliberative action.* Oxford University Press.

Curato, N. (2020). After disinformation: Three experiments in democratic renewal in the Philippines and around the world. In Chua, Y., Curato, N. & Ong, J. (Eds.), *Information dystopia and Philippine democracy* (pp. 76–82). Internews. https://internews.org/sites/default/files/2021-02/Internews_report_information_dystopia_Philippine_democracy_2021-01-updated.pdf

DeLuca, K. M., Lawson, S., & Sun, Y. (2012). Occupy wall street on the public screens of social media: The many framings of the birth of a protest movement. *Communication, Culture & Critique, 5* (4), 483–509. doi:10.1111/j.1753-9137.2012.01141.x

Denisova, A., & Herasimenka, A. (2019). How Russian rap on YouTube advances alternative political deliberation: Hegemony, counter-hegemony, and emerging resistant publics. *Social Media + Society.* doi:10.1177/2056305119835200

Diakopoulos, N. (2015). Algorithmic accountability: Journalistic investigation of computational power structures. *Digital Journalism, 3* (3), 398–415. doi:10.1080/21670811.2014.976411

Diamond, L. (2010). Liberation Technology. *Journal of Democracy, 21* (3), 69–83.

Dylko, I. B., Beam, M. A., Landreville, K. D., & Geidner, N. (2012). Filtering 2008 US presidential election news on YouTube by elites and non-elites: An examination of the democratizing potential of the Internet. *New Media & Society, 14* (5), 832–849. doi:10.1177/1461444811428899

Feroz Khan, G., & Vong, S. (2014). Virality over YouTube: An empirical analysis. *Internet Research, 24* (5), 629–647. doi:10.1108/IntR-05-2013-0085

Gillespie, T. (2018). All platforms moderate. In *Custodians of the Internet: Platforms, content moderation, and the hidden decisions that shape social media* (pp. 7–21). Yale University Press.

GMA News Online (2021). *Eleksyon 2022.* Available at https://www.gmanetwork.com/news/eleksyon2022/ (accessed on 21 May 2022)

Gueorguieva, V. (2008). Voters, MySpace, and YouTube: The impact of alternative communication channels on the 2006 election cycle and beyond. *Social Science Computer Review, 26* (3), 288–300. doi:10.1177/0894439307305636

Hanson, G. L., Haridakis, P. M., & Sharma, R. (2011). Differing uses of YouTube during the 2008 US presidential primary election. *Electronic News, 5* (1), 1–19. doi:10.1177/1931243111398213

Hillmann, H. (2008). Localism and the limits of political brokerage: Evidence from Revolutionary Vermont. *American Journal of Sociology, 114* (2), 287–331. doi:10.1086/590646

Howard, P. N., Duffy, A., Freelon, D., Hussain, M. M., Mari, W., & Maziad, M. (2011). Opening closed regimes: What was the role of social media during the Arab Spring? *SSRN Electronic Journal.* doi:10.2139/ssrn.2595096

Kaplan, A.M., & Haenlein, M. (2010). Users of the world, unite! The challenges and opportunities of social media. *Business Horizons, 53* (1), 59-68. doi:10.1016/j.bushor.2009.09.003

Kerkvliet, B. J. T. (1995). Toward a more comprehensive analysis of Philippine politics: Beyond the patron-client, factional framework. *Journal of Southeast Asian Studies, 26*(02), 401. doi:10.1017/s0022463400007153

Lewis, R. (2018). *Alternative influence: Broadcasting the reactionary right on YouTube.* Data & Society. https://datasociety.net/wp-content/uploads/2018/09/DS_Alternative_Influence.pdf

Loader, B. D., & Mercea, D. (2011). NETWORKING DEMOCRACY? Social media innovations and participatory politics. *Information, Communication & Society, 14* (6), 757–769. doi:10.1080/1369118X.2011.592648

Lobato, R. (2016). The cultural logic of digital intermediaries: YouTube multichannel networks. *Convergence: The International Journal of Research into New Media Technologies, 22* (4), 348–360. doi:10.1177/1354856516641628

Lobato, R., Thomas, J., & Hunter, D. (2012). Histories of user-generated content: Between formal and informal media economies. In D. Hunter, R. Lobato, M. Richardson, & J. Thomas. (Eds.), *Amateur Media: Social, cultural and legal perspectives* (pp. 3–17). Routledge.

Marwick, A.E. (2013). *Status update: Celebrity, publicity, and branding in the social media age.* Yale University Press.

Mercado, N.A (2021). Pulse Asia's December survey: Marcos and Duterte-Carpio team is top choice. Retrieved 8 November 2021, https://newsinfo.inquirer.net/1530873/pulse-asias-december-survey-marcos-and-duterte-carpio-team-is-top-choice#ixzz7KoiuHwwq

McKay, D. (2020). Decorated Duterte: Digital objects and the crisis of Martial Law history in the Philippines. *Modern Languages Open*, (1), p. 27. doi:10.3828/mlo.v0i0.316

Mijares, P. (1976). *The conjugal dictatorship of Ferdinand and Imelda Marcos.* Union Square Publications.

Nowak, T.C., & Snyder, K. (1974). Clientelist politics in the Philippines: Integration or instability? *American Political Science Review, 68* (3), 1147–1170. doi:10.2307/1959153

Official Gazette. (2015). The fall of the dictatorship. Retrieved 10 October 2020, https://www.officialgazette.gov.ph/featured/the-fall-of-the-dictatorship/

Ong, J., & Cabañes, J.V. (2018). Architects of networked disinformation: Behind the scenes of troll accounts and fake news production in the Philippines. *Newton Tech4Dev Network.* doi:10.7275/2cq4-5396

Ong, J. & Cabañes, J.V. (2019). When disinformation studies meets production studies: Social identities and moral justifications in the political trolling industry. *International Journal of Communication, 13,* 5771–5790. https://scholarworks.umass.edu/communication_faculty_pubs/110

Ong, J., Tapsell, R., & Curato, N. (2019). Tracking digital disinformation in the 2019 Philippine midterm election. *New Mandala.* www.newmandala.org/disinformation

Rafael, V.L. (2000). *White love and other events in Filipino history.* Duke University Press.

Rafael, V.L. (2003). The cell phone and the crowd: Messianic politics in the contemporary Philippines. *Public Culture 15* (3), 399–425. https://www.muse.jhu.edu/article/47187

Reyes, M.P. (2018). Producing Ferdinand E. Marcos, the scholarly author. *Philippine Studies: Historical and Ethnographic Viewpoints* 66(2): 173–218.

Reyes, M.P. & Ariate, J. (2020, February 24). 'Marcos truths': A genealogy of historical distortions. [Conference presentation]. Democracy and Disinformation National Conference, Manila, Philippines.

Reyes, M.P. & Jose, R. (2012–2013). Why Marcos *pa rin. Kasarinlan: Philippine Journal of Third World Studies* 27(1–2)-28(1–2): 1–18.

Ressa, M (2016). Part 1, Weaponising the Internet. Retrieved 14 September 2020, www.rappler.com/nation/148007-propaganda-war-weaponizing-internet

Ridout, T. N., Franklin Fowler, E., & Branstetter, J. (2010, August 23). Political advertising in the 21st century: The rise of the YouTube ad [Paper presentation]. American Political Science Association (APSA) 2010 Annual Meeting Paper, Washington, DC. https://ssrn.com/abstract=1642853

Rogers, R. (2013). End of the virtual: Digital methods. In *Digital methods* (pp. 19–38). The MIT Press.

Sandvig, C., Hamilton, K., Karahalios, K., & Langbort, C. (2014, May 22). Auditing algorithms: Research methods for detecting discrimination on Internet platforms [Paper presentation]. Data and discrimination: Converting critical concerns into productive inquiry. 64th Annual Meeting of the International Communication Association, Washington, DC.

Scott, J. (1972). Patron-Client politics and political change in Southeast Asia. *American Political Science Review, 66* (1), 91–113. doi:10.2307/1959280

Senft, T.M. (2013). Microcelebrity and the branded self. In J. Hartley, J. Burgess, & A. Bruns (Eds.), *A Companion to New Media Dynamics* (pp. 346–354). Wiley-Blackwell Publishing.

Soriano, C. R. and Gaw, F. (2021). Platforms, alternative influence, and networked political brokerage on YouTube. *Convergence*. doi:10.1177/13548565211029769

Stovel, K., & Shaw, L. (2012). Brokerage. *Annual Review of Sociology 38*(1), 139–158. doi:10.1146/annurev-soc-081309-150054

Towncr, T. L., & Dulio, D. A. (2011). An experiment of campaign effects during the YouTube election. New Media & Society, 13(4), 626–644. doi:10.1177/1461444810377917

Uldam, J. & Askanius, A. (2013). Online civic cultures? Debating climate change activism on YouTube. *International Journal of Communication, 7*, 1185–1204. https://ijoc.org/index.php/ijoc/article/view/1755

van Dijck, J., & Poell, T. (2013). Understanding social media logic. *Media and Communication, 1* (1), 2–14. doi:10.17645/mac.v1i1.70

Webb, A. & Curato, N. (2019). Populism in the Philippines. In D. Stockemer (Ed.), *Populism around the world: A comparative perspective* (pp. 49–65). Springer. doi:10.1007/978-3-319-96758-5_4

Videos cited

Allen ReacTV. (2019, April 12). *MARCOS GOLD: SEKRETO SA LIKOD NG 192,000 TONS OF GOLD* [Video]. YouTube. https://www.YouTube.com/watch?v=tVu-9IYn1Oo

birador HQ. (2020, March 15). *ANG MGA NAGAWA NI PNOY SA ATING BANSA, MAS TAMANG TANUNG PALA EH MAY NAGAWA NGA BA SYA? #birador* [Video]. YouTube. https://www.YouTube.com/watch?v=7yjS_hiEgN4

Filipino Future. (2020, March 19). *Mga Totoong Dahilan Bakit Ayaw Nila Maging Presidente Si Bongbong Marcos* [Video]. YouTube. https://www.YouTube.com/watch?v=AJeaLQQO6Kw

JDBros. (2017, 31 August). MARCOS GOLD CAN SAVE THE WORLD 987 BILLION DOLLARS AND MILLION TONS OF GOLD IN THE PHILIPPINES [Video]. YouTube. https://www.YouTube.com/watch?v=1mb1Pte0Ob0&t=48s

Jevara PH. (2019, September 12). *Mga Magandang Nagawa ni Ferdinand Marcos | Jevara PH* [Video]. YouTube. https://www.YouTube.com/watch?v=N8cmrcFSPOo

OHJAYCEE. (2020, January 21). *Ang mga HULING SALITA ni Ferdinand Marcos* [Video]. YouTube. https://www.YouTube.com/watch?v=koPc4vsulUM

Sangkay JanJan TV. (2019, March 21). MAY NAABUSO BA TALAGA NOONG MARTIAL LAW? [Video] https://www.YouTube.com/watch?v=HT_AMSRokKk

Sangkay Janjan TV. (2019, July 1). *KUNG NATULOY ANG PLANO NI MARCOS, HIGIT PA SA AMERIKA ANG PILIPINAS* [Video]. YouTube. https://www.YouTube.com/watch?v=ZaTbvBdYVUg

Tinig PH. (2020, February 13). *Pinoy Trivia "Alam mo ba?" Paano Namatay Ang Mga Pangulo Ng Pilipinas* [Video]. YouTube. https://www.YouTube.com/watch?v=5Tf6hbVG7vI

Toto Bee. (2018, February 27). *BAKIT PINA ALIS SI MARCOS mini documentary* [Video]. YouTube. https://www.YouTube.com/watch?v=5Y_MW1yHHro

7   YouTube and Beyond

Abstract

This chapter summarises the book's key arguments and contentions on Philippine digital cultures, foregrounded by the brokering of feminised subjectivity, intimate relations, world-class labour, and partisan politics on YouTube. It recaps the study's proposed conceptual frame of digital brokering, as well as its key dimensions, reiterating the operations and implications of digital brokering in the everyday digital lives of Filipinos within postcolonial and neoliberal spheres. This final chapter leaves the readers with some future research directions, including ways of approaching and investigating YouTubing as a form of platformisation of everyday life in the Global South. It provides a space for critical reflection on understanding the impact and implications of emerging and rapidly evolving platforms on the intimate, informal, and everyday lives of individuals who constantly negotiate existing structural inequalities in a global and digital society.

Keywords: digital brokerage, affective aspiration, neoliberal globalisation, paradoxical reconfiguration, Global South, YouTube

Social media platforms have been pivotal in redefining the conduct of contemporary society. More specifically, we have shown in the previous chapters the role of YouTube not just as a space for people's performance of everyday life, but as a platform where diverse social transactions and aspirations of the Filipino people are enacted, curated, commodified, and brokered. Taking the case of the Philippines, one of the tech-savviest countries in the world, this book unpacks how Filipino YouTubers and YouTube represent and mediate subjectivity, intimate relations, labour practices, as well as personalised political agenda. It also presents a critical stance in approaching YouTube as a platform that is constitutive in the brokering of profitable and relatable persona, lifestyle imaginaries, social

Soriano, Cheryll Ruth and Earvin Charles Cabalquinto: *Philippine Digital Cultures: Brokerage Dynamics on YouTube*. Amsterdam: Amsterdam University Press, 2022
DOI: 10.5117/9789463722445_CH07

mobility tactics, as well as sustained political affiliation in a neoliberal and postcolonial sphere.

As we have presented, no longer just a passive audience, users are enabled by the contemporary media environment to produce content in a DIY fashion or alternatively curate the media that they wish to consume. Celebrating creativity in content creation and discovery through shared knowledge and participation, this pervasive culture ushered by social media has also smoothed the way for micro-celebrities to emerge with their own networks of followers, many of whom are active on YouTube (Burgess & Green, 2018). This participatory environment also comes at a time when forces that promote distrust in traditional media institutions are active, compelling "people formerly known as audiences" (Rosen, 2012) to consume beyond traditional and institutional sources of information and entertainment and participate in the creation and circulation of content. Nonetheless, as illuminated across our data chapters, YouTubers' ordinariness and relatability attract audiences by offering "wisdom from lived experience" that highlight how their brand of content represents those that may be intently or inadvertently screened out by the usual gatekeepers.

We situate the study within the context of postcolonialism and neo-liberalism. For the former, we highlight how colonialist structures have shaped online performativity and platform-based tactics, re-shaping ways of enacting selfhood, as well as imaginaries of identity, intimacy, labour, and political governance. For the latter, given the impact of colonialism and globalisation on Philippine economic policies that often favoured international and business demands and reinforced socio-economic in-equalities, ordinary Filipinos are left scrambling for access to social welfare benefits and opportunities, often through the help of intermediaries and informal channels. Precarious living conditions are addressed by individuals venturing into overseas work, platform labour, or content production on YouTube. By attending to these contexts, we unpack how emerging social transactions in social media do not exist in isolation from its historical, cultural, and ideological categories. This context offers a critical vantage point for analysing the impact of digitalisation on the everyday life of individuals who remain neglected and even exploited by the nation-state in a globalising and networked economy. This is salient in Toyama's (2011) argument that technologies are not transformative nor democratising in themselves, but rather, they amplify human forces along with existing inequalities and missing institutional capacities.

In this final chapter, we highlight the key contributions of the research study in understanding the implications of ubiquitous digital platforms in mediating the digital lives of Filipinos within postcolonial and neoliberal spheres. The discussion is divided into three sections. The first section recaps the conceptual frames of the study. It charts the key frames that have been deployed to investigate the operations and implications of brokerage in a digital space, also attending to how the dynamics of brokerage of digital cultures on YouTube facilitates affective aspirations and paradoxical reconfigurations that encapsulate the strategic performances, community-building strategies, and contradictory outcomes on YouTube. The second section discusses YouTube as a platform and a broker in a highly networked and commodified environment. The chapter concludes by offering some recommendations and future research directions. We emphasise the need to unravel the influences of a web of interconnected factors – socio-historical, economic, political, and technological – in shaping the utility and operation of an online platform. By doing so, a critical lens is proposed, emphasising how the conduct of research on the intersections of digital media, influencer culture, and labour on social media can expose both possibilities underlying marginality in a digital world.

Brokerage as Lens

In Chapter 2, we characterised brokerage as a process through which actors intermediate the flows of goods, information, and knowledge that bridge the gap between individuals and social institutions. We then extend this characterisation of brokerage in a digital context, pinpointing how online users, platforms, and information enact relationships and forge connections via exchanges, interpretation, and consumption of information. As described in case study Chapters 3–6, Filipino YouTubers act as brokers who mediate notions of beauty standards, ideal interracial relationships, world-class labour, and progressive governance.

We approach digital brokering as tied to social, economic, political, and historical frames. The emplacement of brokerage in socio-historical analysis is crucial because this allows us to contextualise aspirations and imaginaries which brokers draw from to generate value for their brokerage role. This means that online performativity is reflective of people's everyday living conditions and negotiations in a postcolonial and neoliberal state. On the one hand, performing gender, intimacies, labour, and politics presents opportunities for identity formation, agentic expression, and community building. On the other hand, these same practices are symptomatic of colonialist legacies. Practices such as cultural whitening, curating interracial intimacies, servicing the global market as English proficient and flexible workers, and associating "national development" with colonialist governance are key examples. Furthermore, the diversity, visibility, and creativity of online performances are incorporated within and co-opted for digital and neoliberal market systems. As a means to navigate informal, digital and precarious environments, online content creators orchestrate and deliver performances that appeal to what's familiar, relatable, palatable, "controversial," and even misleading to harbour likes, shares, and subscriptions, which metrics are converted to profit and data governance.

Our analysis has shown how online narratives have become integral to the formation of "entertainment publics" in the Philippine setting (A. Pertierra, 2021). Our investigation shows how micro-celebrities or influencers deploy a range of branding strategies that allow them to generate and convert cultural, economic, and political capitals. We can see that a distinct characteristic of entertainment publics is the "mirroring" of the strategies deployed by Philippine television, as well as the politics surrounding it, in digital environments. On YouTube, performances may range from the serious, funny, creative, domesticated, and politically charged. We argue that the appeal of such contents is a result of their familiarity and relatability, which typically mimic performances and contents in Philippine television. Here, YouTubers appropriate contents and styles of traditional media, including texts, visuals, music, engagement styles, and so forth. However, it is also through the process of appropriation that representational politics are reinforced, such as highlighting standards of femininity, the positionality of oneself in establishing a flourishing interracial relationship, strategic labour practices, and partisan politics while engaging the unique affordances of social media platforms such as YouTube. In a digital space, these "familiar" narratives mobilise the metrics game, profit making, and formations of capital.

Key Dimensions of Brokerage

Across the case study chapters, we have developed the four frames that guided our investigation, including affective content, discursive style, credibility-building, and platform-specific strategies. We presented how *affective aspirations* are produced and curated on YouTube. We define affective aspirations as a postcolonial and neoliberal dreamwork mobilised through contents, performances, and strategies that stir intimate connections between the YouTubers, the platform, and the viewers. Individuals who consume a range of contents are invited to access and live vicariously in the mediated persona's world of the YouTuber. The YouTuber, as a broker, positions oneself as an "ordinary expert" whose life has been transformed via a range of offline and online knowledge, practices, and tactics. At the other side of the communication channel, the viewers partake in the typically personalised and emotionally charged narratives of the YouTuber, which are translated into clicks, views, and shares. The appeal of the affective contents does not only lie in their aesthetics and language. They attract eyeballs, clicks, and shares as a result of the imaginaries and possibilities they curate and offer, informally and indirectly, among online users who navigate the similar precarious environment they navigate. By examining the narratives embedded in content, we illuminate the desires and aspirations enabled and bridged by content reference, cultures, and practices by Filipino YouTubers. The narratives embodied in YouTubers' content are situated in the broader social, economic, political, and historical conditions in the Philippines. We emphasise that this involves, first, *bridging information that attends to key aspirations* of Filipinos; second, illustrating how these aspirations can be achieved through *relatable experience*; and third, visualising the results of achieving those aspirations through *personal and collective evidence.* Whether this is about how one can achieve a flawless underarm, how one can find a foreign partner online, or earn millions from a digital freelance job, brokers, through their videos and channels, put together relatable and useful information strategically that serve as a map for how their viewers can achieve or at least imagine the possibility for achieving their aspirations. Stories and personalised narratives are presented to curate a desirable self, a meaningful and happy relationship, globally competitive labour, and progressive political stance. More importantly, the multitude of online information indicates the politics of brokering the Filipino aspiration. Whether to have a whiter skin, to find a foreign husband, to find a digital job, or to experience national progress, these aspirations are constructed through a history of postcolonialism and shortcomings in public institutions to address the population's desires.

As we have shown, authentic communication among YouTubers implies an attempt to balance the potential of gains and conveying one's sincerity, whether via a genuine intent to help or an altruistic commitment to "inspire." Whether paid endorsers or not, the brokers' styles make transactions less overt. Authenticity supports and creates credibility in the content they produce and brings them closer to their communities while helping attract new followers and monetising views in return. More importantly, they authenticate and celebrate their everyday life, their own struggles of "finding out for themselves," and showcase life transformations and successes. For instance, we have presented a range of vignettes that emphasise the "journey" of women in whitening and smoothening their underarm or even finding their "special someone." Moreover, we have showcased how YouTubers venture into, navigate and negotiate economic and political landscapes. These contents – rendered across videos and related channels – construct a composite of information that is *relatable* to the publics that they are targeting because they "tickle" people's aspirations and imagination. The platform's affordances of facilitating search and consequently, the visibility of relevant content that one seeks, contribute to this. Importantly, the affirmation drawn from a community of other subscribers and viewers possessing the same aspirations and achieving them, serve to acknowledge, affirm, and reinforce these representations.

A deep dive into the discursive styles, community-building mechanisms, and platform specific strategies on YouTube has allowed us to articulate how a full grasp of the platform vernacular and cultures of use (Rieder et al., 2018) allows YouTube brokers to not just put together disparate information that feed Filipinos' multiple aspirations, but to more effectively convey them. Content creators adopt various styles – texts, visuals, audio, and overlay stickers and graphics – to craft affective and relatable narratives. In some cases, the tactics to build affinity and intimacy with the viewers include the use of local terms pertaining to a collective unit, such as *mga sis* (sisters), *mga kababayan* (my co-Filipinos), or *mga ka-sangkay* (pertaining to a circle of friends). They use affective language that invoke patriotic and nationalistic feelings or a sense of self-worth. This line of messaging makes the information that they share more palatable to ordinary Filipinos.

Nevertheless, online content creators capture people's attention by producing affective content that are also intertwined with the platform's attention logics and constantly evolving features, now including YouTube live. Brokers actively and constantly call the attention of their viewers to comment, like and share the videos through "shout-outs." Brokers elevate this one-off engagement with the users by inviting them to subscribe and be notified ("Press the bell button!") of new videos, as well as to see past videos that they produced.

Online content creators also harness the interactive interface of YouTube to create close relationships and build a sense of community with their viewers. As we have presented in Chapter 2, this function in the brokerage process is crucial because people's capacity to appreciate the information put together by brokers is hinged on how brokers create a vision of imagined social or political connections or actualise these by connecting previously unconnected actors. We observed that brokers are successful when they are able to create "communities" of viewers and subscribers who generate new resources for each other. Brokers engage in a creative process where they actively solicit related topics for next videos or actively invite viewers to share their experiences and success stories in the comments section. These strategies allow them to personalise information and involve the viewers to generate a loop of exchanges and engagement, sometimes also utilised as topics for succeeding videos. Through this process, viewers share and comment on each other's experiences and discover similarities in context, thereby creating some sort of a parasocial community that functions to reinforce their desires and imaginaries of the possibility that these desires can in fact be achieved. The interactions developed allow brokers and their audiences to manufacture sensibility and intimacy (Bucher, 2018) that work to strengthen the resonance of aspirations that they sell through their videos.

Lastly, we unpack how credibility building is harnessed by YouTubers to create value for their content. Here, they highlight their journey and their unique transformations. YouTubers affirm the credibility of the ideas that they share by visualising the progression of how they attain their personal, familial, and professional goals. The visual orientation of the platform allows for these to be magnified – the magical transformation from a dark underarm to a white and smooth underarm, glorious images of a happy wife with a white husband and a blue-eyed baby, screenshots of six-digit pay checks from online freelancing gigs, or footage of national development trickling down to the ordinary Filipino, accompanied by rousing music and attractive visuals. Additionally, they produce "winning stories," providing ample background information often laden with struggle or hardship, building up the narrative before they highlight the evidence of success. These present an image of possibility, an achievable success for aspiring Filipinos with little means, to emulate. As a result, inspirational posts attract other subscribers to share their own experience and reveal their success stories, thereby creating a greater sense of authenticity as these become a collaboration between the broker and their followers – a picture of real-life stories of achieving aspirations or overcoming difficulty. This works to establish the brokers' authority and believability within a growing network of subscribers and viewers.

However, we contend that contradictory outcomes emerge. We introduce the term *paradoxical reconfigurations* to articulate how these content and practices of digital brokerage on YouTube, upon closer analysis, reflect the invisible hand of postcolonial and neoliberal structures that reinforce hierarchy and divide, even as these create the appearance of self-expression, emancipation, and community generation. In terms of performativity, YouTubers harbour visibility, networked connectivity, and a sense of community, which interactions are necessarily translated into metrics and profit both for the platform and the YouTubers. Yet, they also portray and curate aspirations that achieve relatability as these are strongly hinged on colonialist imaginaries – mobility futures afforded by having a whiter underarm or a foreign partner, partaking in a global labour force despite its exploitative tendencies, or erasing the horrors of past dictatorships while emphasising the nostalgia of a progressive and internationally controlled governance. As Athique argues, encounters that are mediated by the digital carry with them a "functional exchange" (Athique, 2019, p. 9). These transactions, generated from brokering the Filipinos' aspirations, construct norms of being and social relations while allowing the broker to benefit economically from the process. Further, performances and strategies also reinforce the capital accumulation and domination of a global platform, YouTube, which harvests and profits from all the content and engagement that are generated from brokerage. Overall, our investigation reveals how the brokerage of social transactions on YouTube magnifies the continuities – of historical, political, social and even traditional media environments, as well as the contradictions embedded in the simultaneous possibilities for liberation and exploitation (Jordan, 2015) ushered by social media's embedding in the Filipino's everyday life.

YouTube as Broker

Although YouTube facilitates the brokerage function of content creators, we can also consider YouTube itself as a broker. YouTube's affordances and logics, and its cultures of use, facilitate brokerage by visibilising brokers' channels and videos, dictating which videos are recommended next, and building connections between creators, their videos, and audiences in an overabundant social field. Through these functions, YouTube essentially: a) curates disparate information and organises them by the logics of search, content categorisation, and construction of people's "algorithmic identity" (Cheney-Lippold, 2017); and b) it brings together previously unconnected actors. These include YouTubers brokering the same theme of white underarm or interracial intimacy that appear in algorithmic recommendations, as well as subscribers and viewers interested in the same content and who come together to reinforce each other's aspirations or political agendas and identifying concrete strategies on how these aspirations can be achieved.

Bucher (2018) argues that platform logics do not necessarily dictate which becomes visible or not, but instead sets conditions through which visibility is constructed. As we have shown through brokers' tactics, they are well attuned to YouTube's logics to achieve visibility in this space. For example, brokers engage in strategies that make them "algorithmically recognizable" (Gillespie, 2017), such as using keywords and tags to respond to present and future search demands, positioning themselves in spaces of interest through categorisation, and sustaining viewer attention by drawing from micro-celebrity engagement tactics. The ability of the brokers to embody and practice the platform vernaculars (Rieder et al., 2017) allows them to gain traction on the platform. YouTube functions as a device that renders particular forms of discourses as "sensible," which sets the "horizons and modalities of what is visible and audible as well as what can be said, thought, made, or done" (Ranciere in Bucher, 2018, p. 68).

Closely related to visibility is the platform's ordering and organising of content which effectively demarcates which narratives are deemed more relevant than others. This ranking culture builds on practices in search engine optimisation, where "quality" can be manufactured by accumulating the appropriate keywords, volume of views, and strategic "backlinks" or connections with other channels (Muller, 2020), which the brokers know well to appropriate.

YouTube's affordance of assigning categories to videos and channels separately allow for the brokers to place themselves within a wide range of categories, such as those relevant to the search keywords [Marcos history, LDR (long distance relationship), *pampaputi ng kili-kili* (underarm whitening), freelancing etc.], and those that appeal to broader audiences ("educational," "entertainment," or "lifestyle"). YouTube relies on the brokers' use of creative labels and tags for their videos. So when one starts to view a video on interracial marriage or underarm whitening, the platform begins to recommend videos of similar content. Traditional brokers fulfil the same role of matching diffused and distant actors and information (Stovel & Shaw, 2012), and so does YouTube. In effect, YouTube enforces these categories at the same time that it obfuscates them, particularly in a cultural milieu where categories like "news" or "entertainment" are made ambiguous by participatory culture (Madden, Lenhart & Fontaine, 2018).

Finally, YouTube facilitates brokerage through its capacity to curate and personalise information and experiences. The process of curation and personalisation begins with the assumptions that the platform makes about the users' interests and preferences, which materialises as their algorithmic identity (Cheney-Lippold, 2017, p. 165). As a new kind of socio-technical broker, not only is it involved in the exchange of values between parties but also in the cultivation of values through the selective exposure of viewers to particular sets of brokers (Soriano & Gaw, 2021). This engenders the convergence of influencers and the platform, with the brokers building on, complementing, or magnifying their narratives through the use of the platform's affordances. These data points are fed content based on their capability to consume. In other words, videos are delivered following marketing logics by recommending the brokers and videos that would keep viewers watching, while constantly collecting more data about what the users watch, who they engaged with, and to which channels they subscribed. As such, YouTube not only amplifies the brokers and their messages, but reinforces their content. Influencers take advantage of this manufactured proximity by building intimacy with users through specific kinds of affective engagement and communication, which creates data-driven parasocial relations between them and their viewers. Reinforced by the platform's governance mechanisms, brokers eliminate the social distance between them and their viewers, as well as facilitate cohesion in the gaps in information and knowledge in the broader political and social structure (Stovel & Shaw, 2012).

Implications and Recommendations

YouTube has become an integral part of Philippine society. From functioning as a space for archiving personalised and intimate encounters (Strangelove, 2010), it has transformed into a commodified and political platform (Burgess & Green, 2009; van Dijck, 2013), offering individual users the opportunity to forge and access a range of capitals – cultural, symbolic, economic, and political. By critically reflecting on the practices, representations, and strategies of Filipino YouTubers, we sketch the study's key implications and offer some future research recommendations.

YouTube showcases the rise of "new" arbiters of taste (Maguire, 2014). As we have shown across the chapters, YouTubers produce and curate a range of aspirations on enacting standards of beauty, intimacies, economic stability, and progressive politics that simultaneously illustrate how people renegotiate the legacies and lived experiences of neoliberalism and post-coloniality while at the same time magnifying existing hierarchies and asymmetries. Although some of the representations advanced by YouTubers align with mainstream media tropes, we have shown that YouTubing has allowed ordinary people to gain influence because their performances on YouTube embody affective aspirations that many other Filipinos subscribe to, along with DIY strategies that appear doable and results achievable to the ordinary person. Amid unlimited airtime and the capacity to enact sequences of strategies, transformation, and triumph across videos, they rise as brokers that construct images of possibility for others, further fortified by the platform's participatory environment. In this regard, we propose that future research can further unpack the possibilities and limits of the diverse types of performativity and platform-specific strategies involved in the brokerage of culture, economy, and politics in contemporary digital times. This approach will provide a focal point to further examine and problematise the democratising potential of digital spaces especially among those who are situated in the peripheries of global and national developments.

Our study is built on a growing study about influencer culture. It contributes to this cohort of studies by deploying a postcolonial perspective, articulating how locally situated social transformations are deeply tied to the uneven operations of global economies and colonialist projects (Gajjala, 2013). By mapping the relationship through which local content production is intertwined with the expansive operations and transformations of global and colonialist systems, our study has shown how global hierarchies are both perpetuated and negotiated. YouTubers imagine, articulate and curate the promise, hope, and grandeur of global and colonialist influences and outcomes, including the notion of satisfying and good life by having a whiter complexion or potentially birthing an interracial baby that can allow one to access various capitals in the future (Figure 7.1). Additionally, advancement is imagined through the flows of dollars and "international" clients or reiterating the possibility of economic progress patterned after, if not surpassing, the progress of colonial masters. Yet, these portraits present narratives that often elide the structural inequalities produced by global domination, including racial and gendered hierarchies, labour precarity, and a range of vulnerabilities as well as abuses of partisan political systems. By engaging with this point, we can further uncover the myths (Duffy, 2017) and paradoxical consequences of a digital, postcolonial and entrepreneurial landscape. To enact this is to further expand research interrogating the impacts of local and global relations in the formation of digital practices and transactions.

In our analysis of a range of videos, we unravelled that content typically communicate possibilities for mobility and transformation through the use of goods, services or DIY approaches. The very "grey" nature of these videos – where information blurs with opinion and in the case of politics, experience can blend with propaganda – complicate how they should be distinguished, understood, examined, and made accountable. YouTubers

Figure 7.1   Brokering aspirations in a postcolony through YouTube

Illustration by Maysa Arabit

are not held accountable through the use of disclaimers or suggesting that what's been said and seen are based on experience or "one's own research." In a traditional media landscape where content is regulated through censorship and editorial gatekeeping, the producer follows certain protocols for exercising caution about the content that is shared and its possible implications. However, amid YouTube's porous governance mechanisms, accountability is made ambiguous because performativity is framed on personal experience. Disclaimers therefore become shields for accountability. The sharing of one's experience (e.g. "This is how aspirations can be achieved, look at how I did it!") constitutes the epistemological logic shared and reinforced within a community of subscribers and viewers. Thus, the subtle endorsement of a whitening product through the sharing of experience of using it (i.e. "It worked wonders!") is also shielded from accountability regardless of whether a viewer construes this as a factual endorsement of a product's quality or opinion.

The case is so for brokers promoting how viewers can find foreign husbands from particular dating apps, which is articulated sans the warning about the abuses and risks that have been well-documented. Beyond issues of physical or domestic abuse, the risks of marriage migration include other sources of socio-legal precariousness such as the lack of programs for social integration or incorporation of migrant wives, legal and institutional concerns in relation to transnational divorce, and even return migration of non-citizen mothers and children (Asis, Piper, & Raghuram, 2019), which are rarely included in these YouTubers' aspirational narratives. While the work of marriage migration recruitment agencies and brokers in the Philippines conducted both online and offline is gaining more attention with the passing of new laws, such as Republic Act No. 10906 (s. 2015), the ambiguity of how to deal with the promotion work embedded in experiential narratives on YouTube makes it difficult to discern the accountability of brokers involved in this process.

The same goes for the brokers' promotion of platform labour to many aspiring Filipino freelance workers who are unable to find work opportunities in the local employment market. Brokers would showcase strategies as well as vignettes of success, supported by images of new houses or cars as the fruits of their labour, or captured moments of enjoying a nomad life of travelling while working. They cascade imaginaries of success and mobility to their subscribers that also create aspirations for the workers who now have to compete with thousands of new aspirants within a generally precarious platform labour market. In effect, they essentially function as local recruiters for digital platforms. They create the notion that anyone can be like them, successful, entrepreneurial, and well-networked. However, the reality is that the skills, networks and assets that allow them to flexibly navigate across digital labour spaces and opportunities to negotiate with clients may not be as easy to come by for others, who will remain in the general labour pool (Soriano & Cabañes, 2020). Without qualification to their claims, labour brokers establish norms and protocols that both mitigate and perpetuate the precarious condition of digital platform workers.

In the context of politics, political propaganda and disinformation are circulated by brokers while hiding under the veil of opinion-making, experience, and "personal research and discovery." As we have shown in Chapter 6, YouTubers can construct political influence by legitimising particular political narratives as knowledge and attempt to erase a turbulent period in history for political agenda. This political brokerage process breeds legitimisation of content into knowledge without accountability, while allowing the brokers and YouTube as well, to gain economically from the process.

As such, we highlight the importance of future research that can further interrogate the need for transparency of manipulative information circulating through the work of brokers on the platform. This can be achieved by exploring online practices and strategies that tend to undermine a safe, inclusive, and democratic society. Furthermore, YouTube, as an enabler and broker of open and free expression, has to be examined by taking a close scrutiny of its policies that often operate with "minimal liability for what those users say or do" (Lewis, 2018, p. 44). Certainly, the platform has begun to play a bigger role in society beyond a space for self-expression. They are spaces that contribute to the shaping of aspirations, facilitating mobilities and consumption of products, and ultimately, the promotion of norms of a good skin, an ideal relationship, viable work opportunity, a progressive nation, and standards of a good political leader. In this regard, we argue that locating and examining accountability in digital environments should cover the domains of strategic practices enacted by human and non-human brokers, such as the YouTuber and the platform. Nonetheless, as YouTube gains greater traction every day, we highlight the need to research more deeply the content produced and circulating on the platform and their relevance in society, as well as their specific implications for policy.

Further, this book is based on research that examined the content and discursive styles of YouTubers. To be able to study YouTube's brokerage role as a platform, it would be useful to systematically study how the platform lends further legitimacy to brokers through the work of platform recommendation systems. Seeing how the platform connects brokers of the similar kind through an assemblage of recommended videos would illustrate YouTube's role in bringing together previously unconnected actors and fortifying particular narratives that attend to its users' aspirations. Significantly, our investigations and approaches can also be extended and applied in various contexts that seek to problematise the consequences of rapidly evolving online platforms in classed, gendered and racialised digital cultures.

Lastly, while there have been a growing number of studies on how audiences engage with online content (Lange, 2014; Wotanis & McMillan, 2014), future research can explore how postcolonial and digital publics engage with a range of contents, practices and strategies in digital environments. One can map the typology of engagements, including those that affirm, extend, or challenge colonialist legacies mapped onto media texts, bodies, systems, and platforms. It also becomes of interest how YouTube publics perceive such aspirational content, how they come to select and trust the YouTubers that they subscribe to and follow, or how they make sense of fluid communities of audiences and engagements that are forged within these spaces. This analysis can connect to how brokers shape their content and styles to win and sustain that trust. Ultimately, such an approach can rethink how the symbiotic relationship between online content creators, audiences, platforms, and algorithmic systems shapes the production and consumption of discursive texts and practices.

Overall, our study has attempted to contextualise our propositions by looking into diverse practices and strategies of YouTubers in the Philippines, set against the backdrop of postcoloniality and neoliberalism. This approach paves the way for our thinking about and proposition of the formations of affective aspirations and paradoxical reconfigurations. For the former, we highlight the dreamwork occurring in online spaces, and the visibility and curation of aspirations that facilitates affective ties and online engagements. As showcased, contents, practices, and strategies are shaped by broader socio-historical, economic and political factors. As Goldberg (2018, p. 2) argues, in a society brimming with anxiety, the production of affective strategies can function as a political strategy, as anxiety produces specific interventions that aim to resolve the underlying anxieties. YouTube brokers enact immaterial labour – labour that is not normally construed as work – which involves the production of the "cultural content of a commodity" (Andrejevic, 2009, p. 416) including constructing standards of taste, forms of entertainment, norms of intimacy, among others, that nudge others to aspire, feel, and act, or what Hardt (1999, p. 94) calls the affective form of immaterial labour. The ambivalence of this labour condition lies in the fact that content producers do this voluntarily, sometimes within the veneer of playfulness, empowerment, or altruistic intent to inspire. This can be construed as a political strategy amid anxiety, albeit doing this in a private commercial platform that not only conditions interactions within its space, but also well set to gain from the social transactions generated in these "free spaces." Relatedly, the ability to freely subscribe, watch, react, and comment on a broker's video is accompanied by the extraction of user-generated data. Thus, this broker-generated construction of aspirations, intimacy, sociality, and community building also generates surplus value within a capitalist exchange, benefiting the broker economically, but also ultimately, the platform (Delfanti & Arvidsson, 2019, pp. 122–123).

Subscribing to Marginality

YouTube has created opportunities for the rise of brokers: a genre of You-Tubers who cascade imaginaries of the "ideal self," an "ideal relationship," an "ideal opportunity," or an "ideal political environment" to Filipinos. To maintain influence in this environment, brokers also perform multiple kinds of work: creating and sustaining communities, producing and exchanging affective and creative contents, being continually responsive to queries, negotiating visibility and sharing aspects of one's personal life, among others. Brokers establish their brand of expertise and authority and at the same time enact a range of communicative strategies underscored by authentic communication, reciprocity, and algorithmic influence that all create aspirational emulation amongst their viewers. In turn, viewers may establish symbolic and social capital by envisioning clear strategies to potentially achieve their aspirations while also assuming social capital by imagined attachment to the broker and the community of regular followers. Invested with the expectation of future returns that are appropriable and convertible, brokers offer a diverse range of life-coaching videos – from underarm whitening to how to earn millions from freelancing – that allow them to build reputation and trust due to the perceived value that they can offer for helping Filipinos achieve their aspirations.

So where do we go from here? As of this writing, many online platforms continue to emerge, delighting users, communities, audiences, and institutions with the promise of expansive and monetised connectivity, sociality, and political and economic alliances. In a tech-savvy country like the Philippines and with a social fabric that is composed of globally distributed individuals, online spaces have increasingly become a breeding ground for creative digital practices that weave personal, familial and communal aspects of everyday lives. However, it is without a doubt that broader social, political and economic structures influence the quality and dynamics of digital lives. The lack of job opportunities and access to social welfare benefits, as well as the conditions facilitated by a global health pandemic, drive but also complicate heightened digital transactions. Additionally, the current Philippine landscape is a by-product of a long history of colonisation and a divisive governance. By taking these factors into account in investigating practices and representations in online spaces, we interrogate and problematise the democratising potential of digital media practices in postcolonial and neoliberal states.

The popularity of YouTube will remain strong in Philippine society amid the growing communicative ecology of practices and relations that move fluidly from one device and platform to another. This, without a doubt, is bolstered by media influences, social and living conditions, and imaginaries of global connections. Amid the rise of new platforms, Filipinos will be continually attracted to "success stories" of YouTubers, which traditional media channels have been constantly showcasing across different channels too. Reports consistently show how Filipino YouTubers earn money, generate networks, finance and build their own house, become "headliners," and even shift from being a viral sensation to becoming big television stars. As such, in these narratives, YouTube is not just operating as a "YouTube academy" where one learns a range of skills and knowledge. But it is continually positioned as a space where economic opportunities, social mobility and political capital can be visibilised, curated, monetised, brokered, and consumed. Significantly, YouTubing affords a global imaginary especially for postcolonial Philippines. In the opening paragraphs of Chapter 1, we highlighted how YouTube is perceived and even framed as an online pathway for fame and fortune, as evinced in the case of Jake Zyrus, Arnel Pineda, and many more. However, while the glamour, fun and networked sociality remain alluring, one cannot deny the power structures that shape success, popularity, and profit. Here, we emphasise how ordinary individuals capitalise on their skills, networks, and aspirations to navigate a global and colonialist system. By reflecting on this stance, it is thus important to question how fast-evolving online platforms will continually shape the lives especially of those who aspire, desire, and remain hopeful for a better life.

References

Andrejevic, M. (2009). *Exploiting YouTube: Contradictions of user-generated labour.* In P. Snickars & P. Vonderau (Eds.). *The YouTube reader* (pp. 406–423). The National Library of Sweden.

Arvidsson, A. (2007). Creative class or administrative class: On advertising and the 'underground.' *Ephemera*, 29(1), 8–23.

Asis, M., Piper, N., & Raghuram, P. (2019). *From Asia to the world: 'Regional' contributions to global migration research.* Revue européenne des migrations internationales, 35(1–2), 13–37.

Bucher T. (2018). *If...then: Algorithmic power and politics.* Oxford University Press.

Burgess, J, & Green, J. (2018). *YouTube: Online video and participatory culture* (Second ed.). Polity.

Burgess, J., & Green, J. (2009). *YouTube: Online video and participatory culture* (First ed.). Polity.

Cabalquinto, E. C., & Soriano, C. R. (2020). 'Hey, I like ur videos. Super relate!' Locating sisterhood in a postcolonial intimate public on YouTube. *Information, Communication & Society, 23*(6), 892–907. doi:10.1080/1369118X.2020.1751864

Cheney-Lippold, J. (2011). A new algorithmic identity: Soft biopolitics and the modulation of control. *Theory, Culture & Society, 28*(6), 164–181.

Duffy, B.E. (2017). *(Not) getting paid to do what you love: Gender, social media, and aspirational work.* Yale University Press.

Gajjala, R. (2013). *Cyberculture and the subaltern: Weavings of the virtual and real.* Lexington Books.

Goldberg, G. (2018). *Antisocial media: Anxious labour in the digital economy.* New York University Press.

Gillespie, T. (2017). Algorithmically recognizable: Santorum's Google problem, and Google's Santorum problem. *Information, Communication & Society*, 20(1), 63–80. doi:10.1080/1369 118X.2016.1199721

Hardt, M. (1999). Affective labour. *Boundary 26*(2), 89–100.

Jordan, T. (2015). *Information politics: Liberation and exploitation in the digital society.* Pluto Press.

Lange, P. (2014). Commenting on YouTube rants: Perceptions of inappropriateness or civic engagement? *Journal of Pragmatics: An Interdisciplinary Journal of Language Studies, 73*, 53–65. doi:10.1016/j.pragma.2014.07.004

Lewis, R. (2018). *Alternative influence: Broadcasting the reactionary right on YouTube.* Data & Society. https://datasociety.net/wp-content/uploads/2018/09/DS_Alternative_Influence.pdf

Maguire, J. S., & Matthews, J. (2014). Thinking with cultural intermediaries. In J. S. Maguire & J. Matthews (Eds.), *The cultural intermediaries reader.* Sage.

Rieder, B., Matamoros-Fernández, A. & Coromina, Ò. (2018). From ranking algorithms to 'ranking cultures': Investigating the modulation of visibility in YouTube search results. *Convergence, 24*(1), 50–68. doi:10.1177/1354856517736982

Rosen, J. (2012). The people formerly known as the audience. In M. Mandiberg (Ed.), *The social media reader* (pp. 13–16). NYU Press.

Soriano C.R., & Cabañes J.V. (2020). Between 'world class work' and 'proletarianised labour': Digital labour imaginaries in the Global South. In E. Polson, L. Schofield-Clarke, & R. Gajjala (Eds.), *The Routledge companion to media and class* (pp. 213–226). Routledge.

Soriano, C. R. R. & Gaw, F. (2021). Platforms, alternative influence, and networked political brokerage on YouTube. *Convergence: The International Journal of Research into New Media Technologies.* doi:10.1177/13548565211029769

Strangelove, M. (2010). *Watching YouTube: Extraordinary videos by ordinary people.* University of Toronto Press, Scholarly Publishing Division.

Toyama, K. (2011). Technology as amplifier in international development. In: Grudin, J. (Ed.) Proceedings of the 2011 iConference (pp. 75–82). ACM Press. doi:10.1145/1940761.1940772

van Dijck, J. (2013). *The culture of connectivity: A critical history of social media.* Oxford University Press.

Wotanis, L., & McMillan, L. (2014). Performing gender on YouTube: How Jenna Marbles negotiates a hostile online environment. *Feminist Media Studies, 14*(6), 912–928. doi:10.1080/14680777 .2014.882373

Bibliography

Abara, A. C. & Heo, Y. (2013). Resilience and recovery: The Philippine IT-BPO industry during the global crisis. *International Area Studies Review, 16*(2), 160–183.

Abidin, C. (2015). Communicative 💙 intimacies: Influencers and perceived interconnectedness. *Ada: A Journal of Gender, New Media, and Technology, 8,* 1–16. doi:10.7264/N3MW2FFG

Abidin, C. (2017). #familygoals: Family influencers, calibrated amateurism, and justifying young digital labour. *Social Media + Society.* doi:10.1177/2056305117707191

Abidin, C. (2018). *Internet celebrity: Understanding fame online* (First ed.). Bingley, UK: Emerald Publishing Limited.

Abidin, C. (2016). Visibility labour: Engaging with Influencers' fashion brands and #OOTD advertorial campaigns on Instagram. *Media International Australia, 161*(1), 86–100. doi:10.1177/1329878X16665177

Abidin, C., & Brown, M. L. (2019). *Microcelebrity around the globe: Approaches to cultures of Internet fame* (First ed.). Emerald Publishing Limited.

Aguilar, F. J. (2014). *Migration revolution: Philippine nationhood and class relations in a globalized age.* Ateneo de Manila University Press.

Aguilar, F.V., Jr. (2019). Political conjuncture and scholarly disjunctures: Reflections on studies of the Philippine state under Marcos. *Philippine Studies: Historical and Ethnographic Viewpoints, 67* (1), 3–30. doi:10.1353/phs.2019.0004

Airoldi, M., Beraldo, D., & Gandini, A. (2016). Follow the algorithm: An exploratory investigation of music on YouTube. *Poetics, 57,* 1–13. doi:10.1016/j.poetic.2016.05.001

Alexander, J., & Losh, E. (2010). A YouTube of one's own? 'Coming out' videos as rhetorical action. In M. Cooper & C. Pullen (Eds.), *LGBT identity and online new media* (pp. 37– 50). Routledge.

Allen ReacTV. (2019, April 12). *MARCOS GOLD: SEKRETO SA LIKOD NG 192,000 TONS OF GOLD* [Video]. YouTube. https://www.YouTube.com/watch?v=tVu-9IYn1Oo

Altmaier, N., Beraldo, D., Castaldo, M., Jurg, D., Romano, S., Renoldi, M. Smirnova, T., Seweryn, N., & Veivo, L. (2019). *YouTube tracking exposed: Apps and their practices.* Available from https://YouTube.tracking.exposed/trexit/

Amoore, L. (2011). Data derivatives: On the emergence of a security risk calculus for our times. *Theory, Culture & Society, 28*(6), 24–43.

Amy, L. E. (1997). The mail-order bride industry and immigration: Combating immigration fraud. *Indiana Journal of Global Legal Studies, 5*(1), 367–374.

Anderson, B. 1991 [1983]. *Imagined communities: Reflections on the origins and spread of nationalism.* Verso.

Andrejevic, M. (2009). *Exploiting YouTube: Contradictions of user-generated labour.* In P. Snickars & P. Vonderau (Eds.). *The YouTube reader* (pp. 406–423). The National Library of Sweden.

Andrejevic, M. (2007). Ubiquitous computing and the digital enclosure movement. *Media International Australia,* (125), 106–117. doi:10.1177/1329878X0712500112

Appadurai, A. (1990). Disjunction and difference in a global cultural economy. In J. Featherstone (Ed.), *Global culture: Nationalism, globalism and modernity* (pp. 295–310). Sage.

Appadurai, A. (1996). *Modernity at large: Cultural dimensions of globalization.* University of Minnesota Press.

Aquino, K. (2018). *Racism and resistance among the Filipino diaspora: Everyday anti-racism in Australia.* Routledge.

Arnado, J. M. (2019). Cultural whitening, mobility and differentiation: Lived experiences of Filipina wives to white men. *Journal of Ethnic and Migration Studies*, 1–17. doi:10.1080/1369 183X.2019.1696668

Arthurs, J., Drakopoulou, S., & Gandini, A. (2018). Researching YouTube. *Convergence: The International Journal of Research into New Media Technologies*, *24*(1), 3–15. doi:.1177/1354856517737222

Arvidsson, A. (2007). Creative class or administrative class: On advertising and the 'underground.' *Ephemera*, *29*(1), 8–23.

Asis, M. (2017). *The Philippines: Beyond labour migration, toward development and (possibly) return*. Migration Policy Institute. Available at https://www.migrationpolicy.org/article/ philippines-beyond-labour-migration-toward-development-and-possibly-return (accessed 1 March 2021)

Asis, M., Piper, N., & Raghuram, P. (2019). *From Asia to the world: 'Regional' contributions to global migration research*. Revue européenne des migrations internationales, *35*(1–2), 13–37.

Aspinall, E., Davidson, M., Hicken, A., & Weiss, M. (2016). Local machines and vote brokerage in the Philippines. *Contemporary Southeast Asia*, *38*(2), 191–196.

Association of Internet Researchers. (2012). *Ethical decision-making and Internet research: Recommendations from the AoIR ethics working committee* (version 2.0). Retrieved 12 January 2020, https://aoir.org/reports/ethics2.pdf

Association of Internet Researchers. (2019). *Internet research: Ethical guidelines 3.0*. Retrieved 10 January 2020, https://aoir.org/reports/ethics3.pdf

Athique, A. (2019). *Integrated commodities in the digital economy. Media, Culture & Society*, *41*(4), 554–570. doi:10.1177/0163443719861815

Baldo-Cubelo, J. T. (2015). The embodiment of the new woman: Advertisements' mobilization of women's bodies through co-optation of feminist ideologies. *Plaridel: A Philippine Journal of Communication, Media, and Society*, *12*(1), 42–65.

Banet-Weiser, S. (2012). *Authentic TM: The politics and ambivalence in a brand culture*. New York University Press.

Banet-Weiser, S., (2011). Branding the post-feminist self: Girls' video production and YouTube. In M. Kearney (Ed.) *Mediated girlhoods: New explorations of girls' media culture* (pp. 277–294). Routledge.

Banet-Weiser, S. (2015). Keynote Address: Media, markets, gender: Economies of visibility in a neoliberal moment. *Communication Review*, *18*(1), 53–70. doi:10.1080/10714421.2015.996398

Banet-Weiser, S., & Juhasz, A. (2011). Feminist labour in media studies/communication: Is self-branding feminist practice? *International Journal of Communication (19328036)*, *5*, 1768–1775.

Bauman, Z. (2000). *Liquid modernity*. Polity Press.

Bautista, V.F. (2018). The pervert's guide to historical revisionism: Traversing the Marcos fantasy. *Philippine Studies: Historical and Ethnographic Viewpoints*, *66* (3), 273–300. doi:10.1353/ phs.2018.0026

Baym, N. (2015). Connect with your audience! The relational labour of connection. *Communication Review*, *18*(1), 14–22. doi:10.1080/10714421.2015.996401

Baym, N.K. (2010). *Personal connections in the digital age* (First ed.). Polity Press.

Benedicto, B. (2013). Queer space in the ruins of dictatorship architecture. *Social Text*, *31*(4 [117]), 25–47. doi:10.1215/01642472-2348977

Benevenuto, F.; Duarte, F.; Rodrigues, T.; Almeida, V. A.; Almeida, J. M.; and Ross, K. W. (2008). *Understanding video interactions in YouTube*. In Proceedings of the 16th ACM International Conference on Multimedia, MM '08, 761–764. New York, USA: ACM.

Berryman, R., & Kavka, M. (2018). Crying on YouTube: Vlogs, self-exposure and the productivity of negative affect. *Convergence*, *24*(1), 85–98.

Berryman, R., & Kavka, M. (2017). 'I guess a lot of people see me as a big sister or a friend': The role of intimacy in the celebrification of beauty vloggers. *Journal of Gender Studies, 26*(3), 307–320. doi:10.1080/09589236.2017.1288611

Bhabha, H. (1984). Of mimicry and man: The ambivalence of colonial discourse. *Discipleship: A Special Issue on Psychoanalysis, 28*, 125–133.

Bhabha, H. (1994). *The location of culture*. Routledge.

birador HQ. (2020, March 15). *ANG MGA NAGAWA NI PNOY SA ATING BANSA, MAS TAMANG TANUNG PALA EH MAY NAGAWA NGA BA SYA? #birador* [Video]. YouTube. https://www.YouTube.com/watch?v=7yjS_hiEgN4

Bishop, S. (2019). Managing visibility on YouTube through algorithmic gossip. *New Media & Society, 21*(11/12), 2589–2606.

Bourdieu, P. (1984). *A social critique of the judgement of taste*. Routledge.

Bucher T (2018) *If...then: Algorithmic power and politics*. Oxford University Press.

Bucher, T. (2016). Neither black nor box: Ways of knowing algorithms. In S. Kubitschko & A. Kaun (Eds.), in *Innovative methods in media and communication research* (pp. 81–98). Springer International Publishing. doi:10.1007/978-3-319-40700-5_5

Bucher, T. (2012). Want to be on the top? Algorithmic power and the threat of invisibility on Facebook. *New Media & Society, 14*(7), 1164–1180. doi:10.1177/1461444812440159

Burgess, J. (2008). 'All your chocolate rain are belong to us?' In G. Lovink & S. Niederer (Eds.), *Video Vortex reader responses to YouTube* (pp. 101–110). Institute of Network Cultures.

Burgess, J. (2015). From 'Broadcast yourself' to 'Follow your interests': Making over social media. *International Journal of Cultural Studies, 18*(3), 281–285. doi:10.1177/1367877913513684

Burgess, J. (2011). User-created content and everyday cultural practice: Lessons from YouTube. In J. Bennett & N. Strange (Eds.), *Television as digital media* (pp. 311–331). Duke University Press.

Burgess, J. (2007). Vernacular creativity and new media (Doctoral dissertation), Queensland University of Technology, Brisbane, Australia. Retrieved 26 October 2020, http://eprints.qut.edu.au/16378/1/Jean_Burgess_Thesis.pdf

Burgess, J., & Green, J. (2009). *YouTube: Online video and participatory culture* (First ed.). Polity.

Burgess, J, & Green, J. (2018). *YouTube: Online video and participatory culture* (Second ed.). Polity.

Burt, R. (2007). *Brokerage and closure: An introduction to social capital*. Oxford University Press.

Cabalquinto, E. C. (2014). At home elsewhere: The transnational kapamilya imaginary in selected ABS-CBN Station IDs. *Plaridel: A Journal of Philippine Communication, Media and Society, 11*(1), 1–26.

Cabalquinto, E. C. (2018a). 'I have always thought of my family first': An analysis of transnational caregiving among Filipino migrant adult children in Melbourne, Australia *International Journal of Communication, 12*, 4011–4029.

Cabalquinto, E. C. (2018b). 'We're not only here but we're there in spirit': Asymmetrical mobile intimacy and the transnational Filipino family. *Mobile Media & Communication, 6*(2), 1–16.

Cabalquinto, E. C. (2022). *(Im)mobile homes: Family life at a distance in the age of mobile media*. Oxford University Press.

Cabalquinto, E. C., & Wood-Bradley, G. (2020). Migrant platformed subjectivity: Rethinking the mediation of transnational affective economies via digital connectivity services. *International Journal of Cultural Studies, 23*(5), 787–802. doi:10.1177/1367877920918597

Cabalquinto, E. & Soriano, C. R. (2020). 'Hey, I like your videos, super relate!' Locating sisterhood in an online intimate public on YouTube. *Information, Communication, and Society*. doi:10.1080/1369118X.2020.1751864

Cabañes, J. V. (2019). Information and communication technologies and migrant intimacies: The case of Punjabi youth in Manila. *Journal of Ethnic & Migration Studies, 45*(9), 1650–1666. do i:10.1080/1369183X.2018.1453790.

Cabañes, J. V. (2014). Multicultural mediations, developing world realities: Indians, Koreans and Manila's entertainment media. *Media, Culture & Society, 36*(5), 628–643.

Cabañes, J. V., & Acedera, K. (2012). Of mobile phones and mother-fathers: Calls, text messages, and conjugal power relations in mother-away Filipino families. *New Media & Society, 14*(6), 916–930. doi:10.1177/1461444811435397

Cabañes, J. V. & Collantes, C. (2020). Dating apps as digital flyovers: Mobile media and global intimacies in a postcolonial city. In J. V. Cabañes & L. Uy-Tioco (Eds.), *Mobile media and social intimacies in Asia: Reconfiguring local ties and enacting global relationships* (pp. 97–114). Springer.

Camba, A. A. (2012). Religion, disaster, and colonial power in the Spanish Philippines in the sixteenth to seventeenth centuries. *Journal for the Study of Religion, Nature and Culture, 6*(2), 215–231. doi:10.1558/jsrnc.v6i2.215

Camus, M. (2019). 3rd telco rollout starts moving. *Inquirer Business,* Retrieved 12 January 2020, https://business.inquirer.net/280343/3rd-telco-rollout-starts-moving

Casilli, A. (2017). Digital labour studies go global: Towards a digital decolonial turn. *International Journal of Communication, 11,* 3934–3954.

Chang, H.H., & Ahn, S.Y. (2020, May 23). #YouTubeAndrewYang: How a random man became a presidential candidate [Paper presentation]. International Communication Association Virtual Conference.

Cheney-Lippold, J. (2011). A new algorithmic identity: Soft biopolitics and the modulation of control. *Theory, Culture & Society, 28*(6), 164–181.

Cheney-Lippold, J. (2017). *We are data: Algorithms and the making of our digital selves.* New York University Press.

Choy, C. C. (2003). *Empire of care: Nursing and migration in Filipino American history.* Duke University Press.

Chu, D. (2009). Collective behavior in YouTube: A case study of 'bus uncle' online videos. *Asian Journal of Communication, 19*(3), 337–353.

Cohn, J. (2019). *The burden of choice: Recommendations, subversion, and algorithmic culture.* Rutgers University Press.

Commission on Filipino Overseas (2018). Statistical profiles of spouses and other partners of foreign nationals. Retrieved 10 May 2020, from https://cfo.gov.ph/statistics-2/

Constable, N. (2005). A Tale of two marriages: International matchmaking and gendered mobility. In N. Constable (Ed.), *Cross-border marriages: Gender and mobility in transnational Asia* (pp. 166-186). University of Pennsylvania Press.

Constable, N. (2006). Brides, maids, and prostitutes: Reflections on the study of 'trafficked' women. *Portal: Journal of Multidisciplinary International Studies, 3*(2), 1–25. doi:10.5130/portal.v3i2.164

Constable, N. (2012). Correspondence marriages, imagined virtual communities, and countererotics on the Internet. In P. Mankekar & L. Schein (Eds.), *Media, erotics, and transnational Asia* (pp. 111–138). Duke University Press.

Constable, N. (2009). The commodification of intimacy: Marriage, sex, and reproductive labour. *Annual Review of Anthropology, 38,* 49. doi:10.1146/annurev.anthro.37.081407.085133

Constable, N. (2003). *Romance on a global stage: Pen pals, virtual ethnography, and 'mail-order' marriages.* University of California Press.

Couldry, N. (2000). *The place of media power: Pilgrims and witnesses of the media age.* Routledge.

Crisostomo, J. (2020). What we do when we #PrayFor: Communicating post humanitarian solidarity through #PrayForMarawi. *Plaridel: A Philippine Journal of Communication, Media, and Society*, 1–34 Advance online publication. http://www.plarideljournal.org/article/what-we-do-when-we-prayfor-communicating-posthumanitarian-solidarity-through-prayformarawi/

Cunningham, S. (2012). Emergent innovation through the coevolution of informal and formal media economies. *Television and New Media*, *13*(5), 415–430. doi:10.1177/1527476412443091

Cunningham, S., & Craig, D. (2017). Being 'really real' on YouTube: Authenticity, community and brand culture in social media entertainment. *Media International Australia*, *164*, 71–81.

Cunningham, S., Craig, D., & Silver, J. (2016). YouTube, multichannel networks and the accelerated evolution of the new screen ecology. *Convergence: The International Journal of Research into New Media Technologies*, *22*(4), 376–391. doi:10.1177/1354856516641620

Curato, N. (2020). After disinformation. Three experiments in democratic renewal in the Philippines and around the world. In Chua, Y., Curato, N. & Ong, J. (Eds.), *Information dystopia and Philippine democracy* (pp. 76–82). Internews. https://internews.org/sites/default/files/2021-02/Internews_report_information_dystopia_Philippine_democracy_2021-01-updated.pdf

Curato, N. (2019). *Democracy in a time of misery: From spectacular tragedies to deliberative action.* Oxford University Press.

Cris & Danica (2017, September 30). 'Long distance relationship story 2017' [Video], YouTube https://www.youtube.com/watch?v=IhUhOooicgo

Cunanan, J. (2020, July 20). 'Paano Takutin Ang Ka LDR Mo Na Mawala Ka Sa Kanya!/WagMo Hayaan Na I-Take For Granted Ka Ng Ka LDR Mo' [Video], YouTube, https://www.youtube.com/watch?v=6qyZCyeJOkY

David, C. (2013). ICTs in political engagement among youth in the Philippines. *The International Communication Gazette*, *75*(3), 322–337.

David, C., Ong, J., & Legara, E. F. T. (2016). Tweeting supertyphoon Haiyan: Evolving functions of Twitter during and after a disaster event. *PloS one*, *11*(3), e0150190. doi:10.1371/journal.pone.0150190

David, E. (2015). Purple-collar labour: Transgender workers and queer value at global call centers in the Philippines. *Gender & Society*, *29*(2), 169.

David, E. J. R. (2013). *Brown skin, white minds: Filipino-American postcolonial psychology.* Information Age Pub. Inc.

David, E. J. R., & Okazaki, S. (2006). The colonial mentality scale (CMS) for Filipino Americans: Scale construction and psychological implications. *Journal of Counselling Psychology*, *53*(2), 241–252. doi:10.1037/0022-0167.53.2.241

De Guzman, O. (2003). Overseas Filipino Workers, labour circulation in Southeast Asia, and the (mis)management of overseas migration programs. Retrieved 13 March 2013, http://kyotoreview.cseas.kyoto-u.ac.jp/issue/issue3/article_281.html

DeLuca, K. M., Lawson, S., & Sun, Y. (2012). Occupy wall street on the public screens of social media: The many framings of the birth of a protest movement. *Communication, Culture & Critique, 5* (4), 483–509. doi:10.1111/j.1753-9137.2012.01141.x

Del Vecchio, C. (2007). Match-made in cyberspace: How best to regulate the international mail order bride industry *Columbia Journal of Transnational Law, 46*(1), 177–216.

Denisova, A., & Herasimenka, A. (2019). How Russian rap on YouTube advances alternative political deliberation: Hegemony, counter-hegemony, and emerging resistant publics. *Social Media + Society*. doi:10.1177/2056305119835200

Department of Information and Communications Technology. (2017). *National broadband plan: Building infostructures for a digital nation.* Diliman, Quezon City: Department of Information and Communications Technology.

de Peuter, G., Cohen, N.S., & Saraco, F. (2017). The ambivalence of coworking: On the politics of an emerging work practice. *European Journal of Cultural Studies, 20*(6), 687–706. doi:10.1177/1367549417732997

Diakopoulos, N. (2015). Algorithmic accountability: Journalistic investigation of computational power structures. *Digital Journalism, 3*(3), 398–415. doi:10.1080/21670811.2014.976411

Diamond, L. (2010). Liberation technology. *Journal of Democracy, 21*(3), 69–83.

Dixon, A. R., & Telles, E. E. (2017). Skin color and colorism: Global research, concepts, and measurement. *Annual Review of Sociology, 43*, 405–424. doi:10.1146/annurev-soc-060116-053315

Dobson, A. R. S. (2015). Girls' 'pain memes' on YouTube: The production of pain and femininity on a digital network. In S. Baker, B. Robards, & B. Buttigieg (Eds.), *Youth Cultures and Subcultures: Australian Perspectives* (pp. 173–182). Ashgate Publishing Limited.

Dobson, A. S., Robards, B., & Carah, N. (2018). Digital intimate publics and social media: Towards theorising public lives on private platforms. In A. S. Dobson, B. Robards, & N. Carah (Eds.), *Digital Intimate Publics and Social Media* (pp. 3–27). Palgrave Macmillan.

Duffy, B. E. (2017). *(Not) getting paid to do what you love: gender, social media, and aspirational work*. Yale University Press.

Dylko, I. B., Beam, M. A., Landreville, K. D., & Geidner, N. (2012). Filtering 2008 US presidential election news on YouTube by elites and non-elites: An examination of the democratizing potential of the Internet. *New Media & Society, 14*(5), 832–849. doi:10.1177/1461444811428899

Edwards, R., & Tryon, C. (2009). Political video mashups as allegories of citizen empowerment. *First Monday, 14*(10). doi:10.5210/fm.v14i10.2617

Elias, A., Gill, R., & Schraff, C. (2017). Aesthetic labour: Beauty politics in neoliberalism. In A. Elias, R. Gill, & C. Schraff (Eds.), *Aesthetic labour: Beauty politics in neoliberalism* (pp. 3–50). Palgrave Macmillan.

Ellison, N., Heino, R., & Gibbs, J. (2006). Managing impressions online: Self-presentation processes in the online dating environment. *Journal of Computer-Mediated Communication* (2), 415.

Ellwood-Clayton, B. (2005). Texting God: The Lord is my textmate – Folk Catholicism in the cyber Philippines. In K. Nyiri (Ed.), *A sense of place: The global and the local in mobile communication* (pp. 251–265). Passagen.

Esguerra, A. (2019, August 11). 'Paano magpaputi ng kili-kili in less than 3 days (my 3 days journey underarm whitening' [Video], YouTube https://www.youtube.com/watch?v=NLeSpFcptmE

Fabros, A. (2016). *Outsourceable selves: An ethnography of call center work in a global economy of signs and selves*. Ateneo de Manila University Press.

Feroz Khan, G., & Vong, S. (2014). Virality over YouTube: An empirical analysis. *Internet Research, 24*(5), 629–647. doi:10.1108/IntR-05-2013-0085

Filipino Future. (2020, March 19). *Mga Totoong Dahilan Bakit Ayaw Nila Maging Presidente Si Bongbong Marcos* [Video]. YouTube. https://www.YouTube.com/watch?v=AJeaLQQO6Kw

Fisher, M. & Taub, A. (2019, August 11). How YouTube radicalized Brazil. *The New York Times*. https://www.nytimes.com/2019/08/11/world/americas/YouTube-brazil.html

Flick, U. (2011). *Introducing research methodology*. SAGE Publications Ltd.

Flores, C. (2019, March 27). 'Paano pumuti ang kili-kili ko using baby oil' [Video], YouTube https://www.youtube.com/watch?v=mftvX9sCmHg

Florida, R. (2014). *The rise of the creative class – revisited: Revised and expanded*. Basic Books.

Fujita-Rony, D. (2010). History through a postcolonial lens: Reframing Philippine Seattle. *The Pacific Northwest Quarterly, 102*(1), 3–13.

Gajjala, R. (2013). *Cyberculture and the subaltern: Weavings of the virtual and real*. Lexington Books.

Garcia, C. R. (2011). Exploring the realities of Korean-Filipino marriages. Retrieved 12 June 2021, https://news.abs-cbn.com/global-filipino/11/07/11/exploring-realities-korean-filipino-marriages

García-Rapp, F. (2017). Popularity markers on YouTube's attention economy: The case of Bubz-beauty. *Celebrity Studies, 8*(2), 228–245. doi:10.1080/19392397.2016.1242430

García-Rapp, F., & Roca-Cuberes, C. (2017). Being an online celebrity: Norms and expectations of YouTube's beauty community. *First Monday, 22*(7), 1–1. doi:10.5210/fm.v22i7.7788

Giddens, A. (1991). *Modernity and self-identity: Self and society in the late modern age*. Polity Press in association with Blackwell Publishing Ltd.

Gill, R., & Scharff, C. (2011). *New femininities postfeminism, neoliberalism, and subjectivity*. Palgrave Macmillan.

Gillespie, T. (2017). Algorithmically recognizable: Santorum's Google problem, and Google's Santorum problem. *Information, Communication & Society, 20*(1), 63–80. doi:10.1080/1369 118X.2016.1199721

Gillespie, T. (2018). All platforms moderate. In *Custodians of the Internet: Platforms, content moderation, and the hidden decisions that shape social media* (pp. 7–21). Yale University Press.

Gillespie, T. (2010). The politics of 'platforms.' *New Media & Society, 12*(3), 347–364. doi:10.1177/1461444809342738

Gillespie, T. (2014). The relevance of algorithms. In T. Gillespie, P. Boczkowski, & K. Foot (Eds.), Media technologies: Essays on communication, materiality, and society (pp. 167–193). MIT Press.

Gillespie, T. (2016). #trendingistrending: When algorithms become culture. In R. Seyfort & J. Roberge (Eds.), *Algorithmic cultures: Essays on meaning, performance, and new technologies* (pp. 52–75). Routledge.

Glatt, Z. (2017). *The commodification of YouTube Vloggers*. (Masters), University of London, Retrieved 18 June 2020, https://zoeglatt.com/wp-content/uploads/2020/05/Glatt-2017-The-Commodification-of-YouTube-Vloggers.pdf

Glatt, Z., & Banet-Weiser, S. (2021). Productive ambivalence, economies of visibility and the political potential of feminist YouTubers. In S. Cunningham & D. Craig (Eds.), *Creator culture: Studying the social media entertainment industry* (pp. 39–56). New York University Press.

Glenn, E. N. (2009). Consuming lightness: Segmented markets and global capital in the skin-whitening trade. In G. E. N. (Ed.), *Shades of difference: Why skin color matters* (pp. 166–187). Stanford University Press.

Glenn, E. N. (2008). Yearning for lightness: Transnational circuits in the marketing and consumption of skin lighteners. *Gender & Society, 22*(3), 281–302. doi:10.1177/0891243208316089

Global Industry Analysts. (2020). Skin lighteners. Market analysis: Trends and forecasts. *Global Industry Analysts*. Retrieved 21 December 2020, from https://www.strategyr.com/MCP-6140.asp

GMA News Online (2021). *Eleksyon 2022*. Available at https://www.gmanetwork.com/news/eleksyon2022/ (accessed on 21 May 2022)

Goggin, G. (2011). *Global Mobile Media*. Routledge.

Goldberg, G. (2018). *Antisocial media: Anxious labour in the digital economy*. New York University Press.

Gonzalez, V. V., & Rodriguez, R. M. (2003). Filipina.com: Wives, workers and whores on the cyber frontier. In R. Lee & S.-l. C. Wong (Eds.), *Asian America.Net: Ethnicity, nationalism, and cyberspace* (pp. 215–234). Routledge.

Graham, M., Hjorth, I. & Lehdonvirta, V. (2017). Digital labour and development: Impacts of global digital labour platforms and the gig economy on worker livelihoods. *Transfer: European Review of Labour and Research, 23*(2), 135–162. doi:10.1177/1024258916687250

Gueorguieva, V. (2008). Voters, MySpace, and YouTube: The impact of alternative communication channels on the 2006 election cycle and beyond. *Social Science Computer Review, 26* (3), 288–300. doi:10.1177/0894439307305636

Guevarra, A. R. (2010). *Marketing dreams, manufacturing heroes: The transnational labour brokering of Filipino workers.* Rutgers University Press.

Hall, R. (1995). The bleaching syndrome: African Americans' response to cultural domination vis-a-vis skin color. *Journal of Black Studies, 26*(2), 172–184.

Hanson, G. L., Haridakis, P. M., & Sharma, R. (2011). Differing uses of YouTube during the 2008 US presidential primary election. *Electronic News, 5* (1), 1–19. doi:10.1177/1931243111398213

Hardon, A. P. (1992). That drug is hiyang for me: Lay perceptions of the efficacy of drugs in Manila, Philippines. *Central Issues in Anthropology, 10*(1), 86–93. doi:10.1525/cia.1992.10.1.86

Hardt, M. (1999). Affective labour. *Boundary 26*(2), 89–100.

Harris, A. (2009). Introduction: Economies of color. In G. E. N. (Ed.), *Shades of difference: Why skin color matters* (pp. 1–5). Stanford University Press.

Hernandez, A. (2020, August 12). '3 Easy steps Kili Kili (mura, tipid at effective)' [Video], YouTube. https://www.youtube.com/watch?v=h-oOvZ873B0

Hesmondhalgh, D. & Baker, S. (2010). 'A very complicated version of freedom': Conditions and experiences of creative labour in three cultural industries. *Poetics, 38*(1), 4–20. doi:10.1016/j.poetic.2009.10.001

Hillmann, H. (2008). Localism and the limits of political brokerage: Evidence from revolutionary Vermont. *American Journal of Sociology, 114* (2), 287–331. doi:10.1086/590646

Hjorth, L. (2011). It's complicated: A case study of personalisation in an age of social and mobile media. *Communication, Politics & Culture, 44*(1), 45–59. https://search.informit.org/doi/10.3316/informit.127649781513814

Hobbs, M., Owen, S., & Gerber, L. (2017). Liquid love?: Dating apps, sex, relationships and the digital transformation of intimacy. *Journal of Sociology, 53*(2), 271–284. doi:10.3316/informit.407452887319646

Holmes, S. (2017.) 'My anorexia story': Girls constructing narratives of identity on YouTube, *Cultural Studies, 31*(1), 1–23, doi:10.1080/09502386.2016.1138978

Hou, M. (2019). Social media celebrity and the institutionalization of YouTube. *Convergence, 25*(3), 534–553. doi:10.1177/1354856517750368

Howard, P. N., Duffy, A., Freelon, D., Hussain, M. M., Mari, W., & Maziad, M. (2011). Opening closed regimes: What was the role of social media during the Arab Spring? *SSRN Electronic Journal.* doi:10.2139/ssrn.2595096

Hunter, M. (2007). The persistent problem of colorism: Skin tone, status, and inequality. *Sociology Compass, 1*(1), 237–254.

Hunter, M. L. (2011). Buying racial capital: skin-bleaching and cosmetic surgery in a globalized world. *Journal of Pan African Studies, 4*(4), 142.

Hunter, M. L. (2005). *Race, gender, and the politics of skin tone.* Routledge.

Hutchinson, J. (2017). *Cultural intermediaries: Audience participation in media organisations.* Palgrave Macmillan.

IBON Foundation (2017). Contractualization prevails. *Facts and Figures, 7*(7). Retrieved 16 May 2020, https://www.ibon.org/contractualization-prevails-ibon-facts-figures-excerpt/

Idzikowski Vlog (2018, February 3). 'Our LDR story to marriage' [Video], YouTube. https://www.youtube.com/watch?v=4DO8ehfhz74.

Information Technology and E-Commerce Council (2003). ITECC Strategic Road Map: Linking Government Services for OFWs. Retrieved 12 June 2019, www.ncc.gov.ph/files/strat_road-mapReport.pdf

iRona TV (2020, June 14). 'How I met my Korean husband 7 years of long distance relationship' [Video], YouTube. https://www.youtube.com/watch?v=1a_JeI3XUUg

Isabel, L. (2019, November 30). 'Paano pumuti ang kili kili / Tipid kili kili routine' [Video],

YouTube. https://www.youtube.com/watch?v=Q3F4twD8X2A

IT & Business Process Association Philippines (IBPAP) (2020). Recalibration of the Philippine IT-BPM Industry Growth Forecasts for 2020–2022. Retrieved 11 March 2021, https://www.ibpap.org/knowledge-hub/research

ITU. (2020). Mobile cellular subscriptions (2001–2019). Retrieved 18 December 2020, https://www.itu.int/en/ITU-D/Statistics/Pages/stat/default.aspx

Jamerson, T. W. (2019). Race, markets, and digital technologies: Historical and conceptual frameworks. In G. D. Johnson, K. D. Thomas, A. K. Harrison, & S. A. Grier (Eds.), *Race in the marketplace.* (pp. 39–54). Palgrave Macmillan.

Jamieson, K.L. (2018). *Cyberwar: How Russian hackers and trolls helped elect a president: What we don't, can't, and do kno*w. Oxford University Press.

Jancsary, D., Hollerer, M. and Meyer, R. (2016). Critical analysis of visual and multimodal texts. In R. Wodak and M. Meye (Eds.), *Methods of critical discourse studies* (pp. 180–204), 3rd ed. Sage.

JDBros. (2017, 31 August). MARCOS GOLD CAN SAVE THE WORLD 987 BILLION DOLLARS AND MILLION TONS OF GOLD IN THE PHILIPPINES [Video]. YouTube. https://www.YouTube.com/watch?v=1mb1Pte0Ob0&t=48s

Jenkins, H. (2006). *Convergence culture: Where old and new media collide.* New York University Press.

Jenkins, H., Ford, S., & Green, J. (2013). *Spreadable media: Creating value and meaning in a networked culture.* New York University Press.

Jensen, S. (2018). Epilogue: Brokers–pawns, disrupters, assemblers? *Ethnos, 83*(5), 888–891. doi:10.1080/00141844.2017.1362455

Jevara PH. (2019, September 12). *Mga magandang nagawa ni Ferdinand Marcos | Jevara PH* [Video]. YouTube. https://www.YouTube.com/watch?v=N8cmrcFSPOo

Jha, M. R. (2016). *The global beauty industry: Colorism, racism, and the national body.* Routledge.

Jordan, T. (2015). *Information politics: Liberation and exploitation in the digital society.* Pluto Press.

Kanai, A. (2019). *Gender and relatability in digital Culture: Managing affect, intimacy and value.* Palgrave Macmillan.

Kaplan, A.M., & Haenlein, M. (2010). Users of the world, unite! The challenges and opportunities of social media. *Business Horizons, 53*(1), 59–68. doi:10.1016/j.bushor.2009.09.003

Kapunan, R. (1991). Labour-only contractors: New generation of '*cabos.' Philippine Law Journal* 65(5), 326.

Kaufman, R. (1974). The patron-client concept and macro-politics: Prospects and problems. *Comparative Studies in Society and History, 16*(3), 284–308. doi:10.1017/S0010417500012457

Kavka, M. (2008). *Reality television, affect and intimacy: Reality matters.* Palgrave Macmillan.

Kelly, P.F. (2001). The political economy of local labour control in the Philippines. *Economic Geography 77*(1), 1–22. doi:10.2307/3594084

Kerkvliet, B. J. T. (1995). Toward a more comprehensive analysis of Philippine politics: Beyond the patron-client, factional framework. *Journal of Southeast Asian Studies, 26*(02), 401–419. doi:10.1017/s0022463400007153

Kern, A., & Müller-Böker, U. (2015). The middle space of migration: A case study on brokerage and recruitment agencies in Nepal. *Geoforum, 65*, 158–169. doi:10.1016/j.geoforum.2015.07.024

Khamis, S., Ang, L., & Welling, R. (2017). Self-branding, 'micro-celebrity' and the rise of social media influencers. *Celebrity Studies, 8*(2), 191–208.

Kleibert, J. M. (2015). Services-led economic development: Comparing the emergence of the offshore service sector in India and the Philippines. In Lambregts, B., Beerepoot, N., Kloosterman, R.C. (Eds.), *The Local impact of globalization in South and Southeast Asia: Offshore business process outsourcing in services industries* (pp. 29–45). Routledge.

Kumar, S. (2016). YouTube nation: Precarity and agency in India's online video scene. *International Journal of Communication, 10*, 5608–5625.

Kusel, V. I. (2014). Gender disparity, domestic abuse and the mail-order bride industry. *Albany Government Law Review, 7*(1), 166–186.

Labour, J. S. J. (2020). Mobile sexuality: Presentations of young Filipinos in dating apps. *Plaridel: A Philippine Journal of Communication, Media, and Society, 17*(1), 247–278.

Laforteza, E. (2016). *The somatechnics of whiteness and race: Colonialism and mestiza privilege.* Taylor and Francis Routledge.

Lange, P. (2014). Commenting on YouTube rants: Perceptions of inappropriateness or civic engagement? *Journal of Pragmatics: An Interdisciplinary Journal of Language Studies, 73*, 53–65. doi:10.1016/j.pragma.2014.07.004

Lange, P. (2007). Publicly private and privately public: Social networking on YouTube. *Journal of Computer-Mediated Communication, 13*(1), 361–380.

Lange, P. (2009). Videos of affinity on YouTube. In P. Snickars & P. Vonderau (Eds.), *The YouTube Reader* (pp. 7–88). National Library of Sweden.

Lange, P.G. (2019). *Thanks for watching: An anthropological study of video sharing on YouTube.* University Press of Colorado.

Lasco, G., & Hardon, A. P. (2020). Keeping up with the times: Skin-lightening practices among young men in the Philippines. *Culture, Health and Sexuality, 22*(7), 838–853. doi:10.1080/13 691058.2019.1671495

Lash, S. (2007). Power after hegemony: Cultural studies in mutation? *Theory, Culture & Society, 24*(3), 55–78.

Lee, D. R. (1998). Mail fantasy: Global sexual exploitation in the mail-order bride industry and proposed legal solutions. *Asian Law Journal, 5*(1), 139–179.

Lehdonvirta, V. (2016). Algorithms that divide and unite: Delocalisation, identity and collective action in microwork. In J. Flecker (Ed.), *Space, place and global digital work* (pp. 53–80). Palgrave Macmillan.

Lewis, R. (2018). *Alternative influence: Broadcasting the reactionary right on YouTube.* Data & Society. https://datasociety.net/wp-content/uploads/2018/09/DS_Alternative_Influence.pdf

Lewis, R. (2020). 'This is what the news won't show you': YouTube creators and the reactionary politics of micro-celebrity. *Television & New Media, 21*(2), 201–217. doi:10.1177/1527476419879919

Lico, G. (2003). *Edifice complex: Power, myth, and Marcos state architecture.* Ateneo de Manila University Press.

Licoppe, C. (2014). *Location awareness and the social-interactional dynamics of mobile sexual encounters between strangers: The uses of Grindr in the gay male community* [Conference session]. Social Lives of Locative Media, Swinburne University of Technology, Melbourne, Australia.

Lin, W., Lindquist, J., Xiang, B. & Yeoh, B.S.A. (2017). Migration infrastructures and the production of migrant mobilities. *Mobilities, 12*(2), 167–174. doi:10.1080/17450101.2017.1292770

Lindee, K. M. (2007). Love, honor, or control: Domestic violence, trafficking, and the questions of how to regulate the mail-order bride industry. *Columbia Journal of Gender and Law, 16*(2), 551–612.

Lindquist, J., Xiang, B., & Yeoh, B. S. A. (2012). Opening the black box of migration: Brokers, the organization of transnational mobility and the changing political economy in Asia. *Pacific Affairs, 85*(1), 7–19. doi:10.5509/20128517

Loader, B. D., & Mercea, D. (2011). Networking democracy? Social media innovations and participatory politics. *Information, Communication & Society, 14* (6), 757–769. doi:10.1080 /1369118X.2011.592648

Lobato, R. (2016). The cultural logic of digital intermediaries: YouTube multichannel networks. *Convergence 22*(4), 348–360. doi:10.1177/1354856516641628

Lobato, R., Thomas, J., & Hunter, D. (2012). Histories of user-generated content: Between formal and informal media economies. In D. Hunter, R. Lobato, M. Richardson, & J. Thomas (Eds.), *Amateur media: Social, cultural and legal perspectives* (pp. 3–17). Routledge. doi:10.4324/9780203112021

Loomba, A. (1998). *Colonialism/postcolonialism. The new critical idiom.* Routledge.

Lorenzana, J. A. (2016). Mediated recognition: The role of Facebook in identity and social formations of Filipino transnationals in Indian cities. *New Media & Society, 18*(10), 2189–2206. doi:10.1177/1461444816655613

Lorenzana, J.A. (2021). The potency of digital media: Group chats and mediated scandals in the Philippines. *Media International Australia.* doi:10.1177/1329878X21988954

Lorenzana, J. A., & Soriano, C. R. R. (2021). Introduction: The dynamics of digital communication in the Philippines: Legacies and potentials. *Media International Australia, 179*(1), 3–8. doi:10.1177/1329878X211010868

Madianou, M., & Miller, D. (2012). *Migration and new media: Transnational families and polymedia.* Routledge.

Magpantay, M. (2019, May 29). 'Effective at murang pampaputi ng Kili-kili/ DIY Tawas Kalamansi' [Video], YouTube. https://www.youtube.com/watch?v=w8VIf7ep6so

Maguire, J. S., & Matthews, J. (2014). Thinking with cultural intermediaries. In J. S. Maguire & J. Matthews (Eds.), *The cultural intermediaries reader.* Sage.

Malaleysvl (2018, January 5). 'Paano paputiin ang kili-kili natural na paraan' [Video], YouTube. https://www.youtube.com/watch?v=cxHqJ3DlKaI

Malefyt, T. & Morais, R. (2012). *Advertising and anthropology: Ethnographic practice and cultural perspectives.* Berg.

Marwick, A. E. (2015). Instafame: Luxury selfies in the attention economy. *Public Culture, 27*(1), 137–160. doi:10.1215/08992363-2798379

Marwick, A. E. (2013). *Status Update: Celebrity, Publicity, and Branding in the Social Media Age.* Yale University Press.

Mateo, J. (2018). Philippines still world's social media capital. Retrieved 19 December 2018, from https://www.philstar.com/headlines/2018/02/03/1784052/philippines-still-worlds-social-media-capital-study

MaximAys Love (2020, April 25). LDR first meeting to marriage love story | 'LDR no more' [Video], YouTube. https://www.youtube.com/watch?v=gfTYuhSM8IE

McKay, D. (2020). Decorated Duterte: Digital objects and the crisis of Martial Law history in the Philippines. *Modern Languages Open*, (1), 27. doi:10.3828/mlo.v0i0.316

McKay, D (2010). On the face of Facebook: Historical images and personhood in Filipino social networking. *History and Anthropology 21*(4), 483–502.

McKay, D., & Perez, P. (2019). Citizen aid, social media and brokerage after disaster. *Third World Quarterly, 40*(10), 1903–1920. doi:10.1080/01436597.2019.1634470

McKenzie, M.J. (2020). Micro-assets and portfolio management in the new platform economy. *Distinktion: Journal of Social Theory.* doi:10.1080/1600910X.2020.1734847

McRobbie, A. (2004). Post-feminism and popular culture. *Feminist Media Studies, 4*(3), 255–264. doi:10.1080/1468077042000309937

Mercado, N.A (2021). Pulse Asia's December survey: Marcos and Duterte-Carpio team is top choice. Retrieved 8 November 2021, https://newsinfo.inquirer.net/1530873/pulse-asias-december-survey-marcos-and-duterte-carpio-team-is-top-choice#ixzz7KoiuHwwq

Melendez, K. (2019, October 12). 'Kili-kili routine/ Paano pumuti at kuminis ang kili-kili ko' [Video], YouTube. https://www.youtube.com/watch?v=y6WcIqOFdfg

Mendoza, R. L. (2014). The skin whitening industry in the Philippines. *Journal of Public Health Policy, 35*(2), 219–238.

Mercurio, R. (2019). *Philippines among top markets for YouTube.* Retrieved 1 November 2019, https://www.philstar.com/business/2019/07/28/1938388/philippines-among-topmarkets-YouTube

Mijares, P. (1976). *The conjugal dictatorship of Ferdinand and Imelda Marcos I.* Union Square Publications.

Mimi Luarca. (2019). *Be a part-time transcriptionist in REV and earn 11,000 pesos! Work from home English subtitles.* [Video]. Retrieved 12 September 2020, https://www.YouTube.com/watch?v=V6oESxlwowM

Miss Kimmy (2018, November 2). Murang Pampaputi ng kili-kili Super Effective (Sorry sa daldal newbie days! Pls, Read description) [Video], YouTube https://www.youtube.com/watch?v=6PQXgWMCy8Q

Nakamura, L. (2013). *Cybertypes: Race, ethnicity, and identity on the Internet.* Routledge.

Nakamura, L., & Chow-White, P. (2012). *Race after the Internet.* Routledge.

Negus, K. (2002). The work of cultural intermediaries and the enduring distance between production and consumption. *Cultural Studies,* 16(4), 501–515, doi:10.1080/09502380210139089

Neilson, B., & Rossiter, N. (2008). Precarity as political concept, or, Fordism as exception. *Theory, Culture & Society* 25(7–8), 51–72. doi:10.1177/0263276408097796

Nielsen, M. (2016). Love Inc.: Toward structural intersectional analysis of online dating sites and applications. In S. U. Noble & B. M. Tynes (Eds.), *The intersectional Internet: Race, sex, class, and culture online* (pp. 161–178). Peter Lang.

Ng, D., & Lachica, L. (2020). The dangers of trying to be fairest of them all in the Philippine. Retrieved 12 November 2020, https://www.channelnewsasia.com/news/cnainsider/dangers-trying-be-fairest-of-them-all-philippines-skin-whitening-12642522

Nowak, T.C. & Snyder, K.A. (1974). Clientelist politics in the Philippines: Integration or instability? *American Political Science Review* 68(3): 1147–1170. doi:10.2307/1959153

Official Gazette. (2015). The Fall of the Dictatorship. Retrieved 10 October 2020, https://www.officialgazette.gov.ph/featured/the-fall-of-the-dictatorship/

Ofreneo, R.E. (2013). Precarious Philippines: Expanding informal sector, 'flexibilizing' labour market. *American Behavioral Scientist, 57*(4), 420–443. doi:10.1177/0002764212466237

OHJAYCEE. (2020, January 21). *Ang mga HULING SALITA ni Ferdinand Marcos* [Video]. YouTube. https://www.YouTube.com/watch?v=koPc4vsulUM

Olchondra, R. (2011, October 13). YouTube Philippines launched. *Inquirer.* https://technology.inquirer.net/5395/YouTube-philippines-launched

Ong, A. (2006). *Neoliberalism as exception mutations in citizenship and sovereignty.* Duke University Press.

Ong, J. (2015). *The poverty of television: The mediation of suffering in class-divided Philippines.* Anthem Press.

Ong, J. & Cabañes, J. (2018). *Architects of networked disinformation: Behind the scenes of troll accounts and fake news production in the Philippines.* Newton Tech4Dev Network. doi:10.7275/2cq4-5396

Ong, J. & Cabañes, J.V. (2019). When disinformation studies meets production studies: Social identities and moral justifications in the political trolling industry. *International Journal of Communication, 13,* 5771–5790. https://scholarworks.umass.edu/communication_faculty_pubs/110

Ong, J. C., & Combinido, P. (2018). Local aid workers in the digital humanitarian project: Between 'second class citizens' and 'entrepreneurial survivors.' *Critical Asian Studies, 50*(1), 86–102. doi:10.1080/14672715.2017.1401937

Ong, J. C., Tapsell, R., & Curato, N. (2019). *Tracking digital disinformation in the 2019 Philippine midterm election*. Retrieved from https://www.newmandala.org/wp-content/uploads/2019/08/Digital-Disinformation-2019-Midterms.pdf

Ookla (2020). Speedtest global index. Retrieved 14 November 2020, https://www.speedtest.net/global-index

Oreglia, E. & Ling, R. (2018). Popular digital imagination: Grassroot conceptualization of the mobile phone in the Global South. *Journal of Communication, 68*. doi:10.1093/joc/jqy01

Padios, J. (2018). *A Nation on the line: Call centers as postcolonial predicaments in the Philippines*. Duke University Press.

Paragas, F. (2009). Migrant workers and mobile phones: Technological, temporal, and spatial simultaneity In R. S. Ling & S. Campbell (Eds.), *The reconstruction of space and time: Mobile communication practices* (pp. 39–66). Transaction.

Pariser, E. (2011). *The filter bubble: What the Internet is hiding from you*. Penguin Press.

Parreñas, R. S. (2011). *Illicit flirtations: Labour, migration and sex trafficking in Tokyo*. Stanford University Press.

Parreñas, R. S. (2001). *Servants of globalization: Women, migration and domestic work*. Stanford University Press.

Parreñas, R. S. (2008). Transnational fathering: Gendered conflicts, distant disciplining and emotional gaps. *Journal of Ethnic and Migration Studies, 34*(7), 1057–1072. doi:10.1080/13691830802230356

Pasquale, F. (2015). The black box society: The secret algorithms that control money and information. Harvard University Press.

Payoneer (2020). *Freelancing in 2020: An abundance of opportunities*. Retrieved 10 July 2021, https://pubs.payoneer.com/docs/2020-gig-economy-index.pdf

Pertierra, A. (2017). Celebrity politics and televisual melodrama in the age of Duterte. In N. Curato (Ed.), *A Duterte reader: Critical essays on Rodrigo Duterte's early Presidency* (pp. 219–229). Ateneo de Manila University Press.

Pertierra, A. (2021). Entertainment publics in the Philippines. *Media International Australia*, 1–14. doi:10.1177/1329878X20985960

Pertierra, A. (2018). Televisual experiences of poverty and abundance: Entertainment television in the Philippines. *The Australian Journal of Anthropology*, 1, 3. doi:10.1111/taja.12261

Pertierra, R. (2020). Anthropology and the AlDub nation, entertainment as politics and politics as entertainment. *Philippine Studies: Historical & Ethnographic Viewpoints, 64*, 289–300.

Pertierra, R. (2006). *Transforming technologies: Altered selves, mobile phone and Internet use in the Philippines*. De La Salle University Press.

Pertierra, R., Ugarte, E., Pingol, A., Hernandez, J., & Dacanay, N. L. (2002). *TXT-ing selves: cellphones and Philippine modernity*. De La Salle University Press.

Philippine Statistics Authority (PSA) (2018). *2015/2016 Industry Profile: Business Process Outsourcing (First of a series), LabStat Updates*. Retrieved 20 December 2020, https://psa.gov.ph/content/20152016-industry-profile-business-process-outsourcing-first-series-0

PKBY (2016, May 1). 'LDR!!! Best surprise visit ever – couple reunited' [Video], YouTube. https://www.youtube.com/watch?v=TIdKF-tefSA

POEA. (2019). Philippine Overseas Employment Administration, 2015–2016 Overseas Employment Statistics. Retrieved 21 July 2019, www.poea.gov.ph/ofwstat/compendium/2015-2016%20OES%201.pdf

Poell, T., Neiborg, D., & Duffy, B. E. (2022). *Platforms and Cultural Production*. Polity Press.

Quiroz, P. A. (2013). From finding the perfect love online to satellite dating and 'loving-the-one you're near': A look at Grindr, Skout, Plenty of Fish, Meet Moi, Zoosk and Assisted Serendipity. *Humanity & Society, 37*(2), 181–185. doi:10.1177/0160597613481727

Radhakrishnan, S. (2011). *Appropriately Indian: Gender and culture in a new transnational class.* Duke University Press.

Rafael, V. (2000). *White love and other events in Filipino history.* Duke University Press.

Rafael, V.L. (2003). The cell phone and the crowd: Messianic politics in the contemporary Philippines. *Public Culture 15*(3), 399–425. https://www.muse.jhu.edu/article/47187

Raun, T. (2018). Capitalizing intimacy: New subcultural forms of micro-celebrity strategies and affective labour on YouTube. *Convergence, 24*(1), 99–113.

Ressa, M (2016). Part 1, Weaponising the Internet. Retrieved 14 September 2020, www.rappler. com/nation/148007-propaganda-war-weaponizing-internet

Reviglio, U., & Agosti, C. (2020). Thinking outside the black-box: The case for 'algorithmic sovereignty' in social media. *Social Media + Society.* doi:10.1177/2056305120915613

Reyes, M.P. (2018). Producing Ferdinand E. Marcos, the scholarly author. *Philippine Studies: Historical and Ethnographic Viewpoints 66*(2), 173–218.

Reyes, M.P. & Ariate, J. (2020, February 24). 'Marcos truths': A genealogy of historical distortions. [Conference presentation]. Democracy and Disinformation National Conference, Manila, Philippines.

Reyes, M.P. & Jose, R. (2012–2013). Why Marcos *pa rin. Kasarinlan: Philippine Journal of Third World Studies* 27(1–2)-28(1–2): 1–18.

Ridout, T. N., Franklin Fowler, E., & Branstetter, J. (2010, August 23). Political advertising in the 21st century: The rise of the YouTube ad [Paper presentation]. American Political Science Association (APSA) 2010 Annual Meeting Paper, Washington, DC. https://ssrn.com/abstract=1642853

Rieder, B., Matamoros-Fernández, A. & Coromina, Ò. (2018). From ranking algorithms to 'ranking cultures': Investigating the modulation of visibility in YouTube search results. *Convergence* 24(1), 50–68.

Robinson, K. (1996). Of mail-order brides and 'boys' own' tales: representations of Asian-Australian marriages. *Feminist Review*(52), 53–68. doi:10.1057/fr.1996.7

Roces, M. (2009). Prostitution, women's movements and the victim narrative in the Philippines. *Women's Studies International Forum, 32*, 270–280. doi:10.1016/j.wsif.2009.05.012

Rodriguez, R. M. (2010). *Migrants for export: How the Philippine state brokers to the world.* The University of Minnesota Press.

Rogers, R. (2013). End of the virtual: Digital methods. In *Digital methods* (pp. 19–38). The MIT Press. https://www.jstor.org/stable/j.ctt5hhd3c

Rondilla, J. (2009). Filipinos and the color complex, Ideal Asian beauty. In G. E. N. (Ed.), *Shades of difference: Why skin color matters* (pp. 63–80). Stanford University Press.

Rondilla, J. L. (2012). *Colonial faces: Beauty and skin color hierarchy in the Philippines and the U.S.* (Doctoral dissertation), University of California, Berkeley, Berkeley, CA.

Rondina, J. (2004). The e-mail order bride as postcolonial other: Romancing the Filipina in web based narratives. *Plaridel: A Philippine Journal of Communication, Media, and Society, 1*(1), 47–56.

Rondilla, J. L., & Spickard, P. (2007). *Is lighter better?: Skin-tone discrimination among Asian Americans.* Rowman & Littlefield Publishers.

Rosen, J. (2012). The people formerly known as the audience. In M. Mandiberg (Ed.), *The social media reader* (pp. 13–16). New York University Press.

Ruiz, J. T. (2019, November 27). *Online jobs at home Philippines for beginners (full tutorial).* [Video]. YouTube. https://www.YouTube.com/watch?v=GUo_3FZtISA

Saban, L. I. (2015). Entrepreneurial brokers in disaster response network in typhoon Haiyan in the Philippines. *Public Management Review, 17*(10), 1496–1517. doi:10.1080/14719037.201 4.943271

Salamanca, M. (2019, October 4). 'Dark underams? Tips and tricks' [Video], YouTube https://www.youtube.com/watch?v=T2EKd66U1CA.

Sandvig, C., Hamilton, K., Karahalios, K., & Langbort, C. (2014, May 22). Auditing algorithms: Research methods for detecting discrimination on Internet platforms [Paper presentation]. Data and discrimination: Converting critical concerns into productive inquiry. 64th Annual Meeting of the International Communication Association, Washington, DC.

Sangkay Janjan TV. (2019, July 1). *KUNG NATULOY ANG PLANO NI MARCOS, HIGIT PA SA AMERIKA ANG PILIPINAS* [Video]. YouTube. https://www.YouTube.com/watch?v=ZaTbvBdYVUg

Sangkay JanJan TV. (2019, March 21). MAY NA ABUSO BA TALAGA NOONG MARTIAL LAW? [Video] https://www.YouTube.com/watch?v=HT_AMSRokKk

San Juan, E. (2011). Contemporary global capitalism and the challenge of the Filipino diaspora. *Global Society, 25*(1), 7–27. doi:10.1080/13600826.2010.522983

San Juan, E. (2009). Overseas Filipino workers: The making of an Asian-Pacific diaspora. *The Global South, 3*(2), 99–129.

Santos, S. (2019, Aug 12). How to make money on Fiverr without skills/ How to make money on Fiverr for beginners. [Video]. YouTube. https://www.youtube.com/watch?v=otQ2I4WhSCw

Saroca, C. (2007). Filipino women, migration and violence in Australia: Lived reality and media image. *Kasarinlan: Philippine Journal of Third World Studies, 21*(1), 75–110.

Saroca, C. (2002). *Hearing the voices of Filipino women: Violence, media representation and contested realities.* (Doctor of Philosophy), University of Newcastle, Newcastle.

Scott, J. (1972). Patron-client politics and political change in Southeast Asia. *American Political Science Review, 66*(1), 91–113. doi:10.2307/1959280

Senft, T. M. (2008). *Camgirls: Celebrity & community in the age of social networks.* Peter Lang.

Senft, T.M. (2013). Microcelebrity and the branded self. In J. Hartley, J. Burgess, & A. Bruns (Eds.), *A Companion to New Media Dynamics* (pp. 346–354). Wiley-Blackwell Publishing.

Shey (2020, May 2). 'Mura at Mabisang Pampaputi ng kili kili | Kili Kili Routine' [Video], YouTube https://www.youtube.com/watch?v=JLqmNG-8yEk.

Shome, R. & Hegde, R.S. (2002). Postcolonial approaches to communication: Charting the terrain, engaging the intersections. *Communication Theory*, 12, 249–270. doi:10.1111/j.1468-2885.2002. tb00269.x

Shresta, T., & Yeoh, B. S. A. (2018). Introduction: Practices of brokerage and the making of migration infrastructure in Asia. *Pacific Affairs, 91*(4), 663–672. doi:10.5509/2018914663

Shtern, J., Hill, S., & Chan, D. (2019). Social media influence: Performative authenticity and the relational work of audience commodification in the Philippines. *International Journal of Communication (19328036)*, 13, 1939–1958.

Silarde, V.Q. (2020). Historical roots and prospects of ending precarious employment in the Philippines. *Labour and Society* 23(4): 461–484.

Simply Rhaze (2016, June 26). 'Our love story: Storytime,' *YouTube* (Simply Rhaze) [Video], YouTube. https://www.youtube.com/watch?v=vDsW4IZeFqc

Sobande, F. (2019). Constructing and critiquing interracial couples on YouTube. In G. D. Johnson, K. D. Thomas, A. K. Harrison, & S. A. Grier (Eds.), *Race in the marketplace.* (pp. 73–85). Palgrave Macmillan

Sobande, F. (2017). Watching me watching you: Black women in Britain on YouTube. *European Journal of Cultural Studies, 20*(6), 655–671. doi:10.1177/1367549417733001

Soriano, C. R. (2019). Communicative assemblages of the '*pisonet*' and the translocal context of ICT for the 'have-less': Innovation, inclusion, stratification. *International Journal of Communication.* https://ijoc.org/index.php/ijoc/article/view/10931

Soriano, C. R. (2021). Digital labour in the Philippines: Emerging forms of brokerage. *Media International Australia.* doi:10.1177/1329878X21993114

Soriano, C. R., & Cabañes, J. V. (2020a). Between 'world-class work' and 'proletarianized labour': Digital labour imaginaries in the global South. In E. Polson, L. S. Clark, & R. Gajjala (Eds.), *The Routledge companion to media and class* (pp. 213–226). Routledge.

Soriano, C. R., & Cabañes, J. V. (2020b). Entrepreneurial solidarities: Social media collectives and Filipino digital platform workers. *Social Media + Society, 6*(2). doi:10.1177/2056305120926484

Soriano, C. R. & Gaw, F. (2021). Platforms, alternative influence, and networked political brokerage on YouTube. *Convergence: The International Journal of Research into New Media Technologies.* doi:10.1177/13548565211029769

Soriano, C. R. & Gaw, M.F. (2020). *Banat by: Broadcasting news against newsmakers on YouTube.* Retrieved 30 July 2020, https://www.rappler.com/voices/imho/analysis-banat-by-broadcasting-news-YouTube-against-newsmakers

Soriano, C. R., Hjorth, L., & Davies, H. (2019). Social surveillance and Let's Play: A regional case study of gaming in Manila slum communities. *New Media and Society, 21*(10), 2119–2139. doi:10.1177/1461444819838497

Soriano, C.R., & Panaligan J.H. (2019). Skill-makers in the platform economy: Transacting digital labour: In A. Athique, & E. Baulch (Eds.), *Digital transactions in Asia: Economic, informational and social exchanges* (pp. 172–191). Routledge. doi:10.33767/osf.io/z4wun

Spivak, G. C. (1999). *A critique of postcolonial reason: Toward a history of the vanishing present.* Harvard University Press.

Starr, E., & Adams, M. (2016). The domestic exotic: Mail-order brides and the paradox of globalized intimacies. *Signs: Journal of Women in Culture & Society, 41*(4), 953–975. doi:10.1086/685480.

Steve Jean Forever (2017, May 4), 'Long Distance Relationship Filipina and American meet for the first time' [Video], YouTube. https://www.youtube.com/watch?v=qFLE6COmVJw

Stockdale, C., & McIntyre, D. (2011). The ten nations where Facebook rules the Internet. Retrieved 12 August 2014, from http://247wallst.com/technology-3/2011/05/09/the-ten-nations-where-facebook-rules-the-internet/3/

Stovel, K., & Shaw, L. (2012). Brokerage. *Annual Review of Sociology, 38*(1), 139–158. doi:10.1146/annurev-soc-081309-150054

Strangelove, M. (2010). *Watching YouTube: Extraordinary videos by ordinary people.* University of Toronto Press, Scholarly Publishing Division.

Strengers, Y., & Nicholls, L. (2017). Aesthetic pleasures and gendered tech-work in the 21st-century smart home. *Media International Australia.* doi:10.1177/1329878X17737661.

Tadiar, N. X. M. (2004). *Fantasy production: Sexual economies and other Philippine consequences for the New World Order.* Hong Kong University Press.

Team TJ (2016, September 7). 'LDR meeting for the first time' [Video], YouTube, https://www.youtube.com/watch?v=uQUKr_aaQhc

Thompson, N. (2018, March 15). Susan Wojcicki on YouTube's fight against misinformation. *Wired.* https://www.wired.com/story/susan-wojcicki-on-YouTubes-fight-against-misinformation/

Thurlow, C., & Jaworski, A. (2017). The discursive production and maintenance of class privilege: Permeable geographies, slippery rhetorics. *Discourse & Society, 28*(5), 535–558. doi:10.1177/0957926517713778

Tinig PH. (2020, February 13). *Pinoy trivia 'Alam mo ba?' Paano namatay ang mga pangulo ng Pilipinas* [Video]. YouTube. https://www.YouTube.com/watch?v=5Tf6hbVG7vI

Tolentino, R. B. (1996). Bodies, letters, catalogs: Filipinas in transnational space. *Social Text*(48), 49–76. doi:10.2307/466786

Torno, S. (2018, August 17). 'Long distance relationship meeting For The First Time!' [Video], YouTube. https://www.youtube.com/watch?v=Me4eUv8Qvrk

Toto Bee. (2018, February 27). *BAKIT PINA ALIS SI MARCOS mini documentary* [Video]. YouTube. https://www.YouTube.com/watch?v=5Y_MW1yHHr0

Towner, T. L., & Dulio, D. A. (2011). An experiment of campaign effects during the YouTube election. *New Media & Society, 13*(4), 626–644. doi:10.1177/1461444810377917

Toyama, K. (2011). Technology as amplifier in international development. In J. Grudin (Ed.) Proceedings of the 2011 iConference (pp. 75–82). ACM Press. doi:10.1145/1940761.1940772

Turner, G. (2009). *Ordinary people and the media: The demotic turn.* SAGE.

Uldam, J. & Askanius, A. (2013). Online civic cultures? Debating climate change activism on YouTube. *International Journal of Communication, 7,* 1185–1204. https://ijoc.org/index.php/ijoc/article/view/1755

UNESCO. (n.d.). Overview of internal migration in the Philippines. Retrieved 22 October 2020, https://bangkok.unesco.org/sites/default/files/assets/article/Social%20and%20Human%20Sciences/publications/philippines.pdf

Utz, S. & Wolfers, L. (2020). How-to videos on YouTube: The role of the instructor. *Information, Communication & Society,* August 2020: 1–16. doi:10.1080/1369118X.2020.1804984

Uy-Tioco, C. (2019). 'Good enough' access: Digital inclusion, social stratification, and the reinforcement of class in the Philippines. *Journal of Communication Research & Practice 5*(2), 156–171.

Uy-Tioco, C. (2007). Overseas Filipino workers and text messaging: Reinventing transnational mothering. *Continuum, 21*(2), 253–265. doi:10.1080/10304310701269081

Uy-Tioco, C., & Cabalquinto, E. C. (2020). Transnational digital carework: Filipino migrants, family intimacy, and mobile media. In J. V. Cabañes & C. Uy-Tioco (Eds.), *Mobile Media and Asian Social Intimacies,* (pp. 153–170). Springer.

van Dijck, J. (2013). *The culture of connectivity: A critical history of social media.* Oxford University Press.

van Dijck, J., & Poell, T. (2013). Understanding social media logic. *Media and Communication, 1*(1), 2–14. doi:10.17645/mac.v1i1.70

Visconti, K. (2012, 19 September). LTE now commercially available in PH. *Rappler.* http://www.rappler.com/business/11169-lte-now-commercially-available-in-ph

Vitorio, R. (2019). Postcolonial performativity in the Philippine heritage tourism industry. In A. Mietzner & A. Storch (Eds.), *Language and tourism in postcolonial settings* (pp. 106–129). Channel View Publications.

We are Social (2020). Digital 2020: Global digital overview. Retrieved 22 November 2021, https://wearesocial.com/digital-2020

We are Social (2021). Digital 2021. Global overview report. Retrieved 20 November 2021, https://wearesocial.com/digital-2021

We are Social (2018). Global digital report 2018. Retrieved 29 November 2020, https://digitalreport.wearesocial.com/download

Webb, A. & Curato, N. (2019). Populism in the Philippines. In D. Stockemer (Ed.), *Populism around the world: A comparative perspective* (pp. 49–65). Springer. doi:10.1007/978-3-319-96758-5_4

Williams, R. (2016). *The influencer economy: How to launch your idea, share it with the world, and thrive in a digital age.* Ryno Lab.

Wilson, C. (2014). The selfiest cities in the world: TIME's definitive ranking. *Time Magazine.* http://time.com/selfies-cities-world-rankings/

Wood, A.J., Lehdonvirta, V., & Graham, M. (2018). Workers of the Internet unite? Online freelancer organisation among remote gig economy workers in six Asian and African countries. *New Technology, Work and Employment, 33*(2), 95–112. doi:10.1111/ntwe.12112

Wotanis, L., & McMillan, L. (2014). Performing gender on YouTube: How Jenna Marbles negotiates a hostile online environment. *Feminist Media Studies, 14*(6), 912–928. doi:10.1080/14680777.2014.882373.

Wyatt, S. (2021). Metaphors in critical Internet and digital media studies. *New Media & Society, 23*(2), 406–416. doi:10.1177/1461444820929324

Yip, J., Ainsworth, S., & Hugh, M. T. (2019). Beyond whiteness: Perspectives on the rise of the Pan-Asian beauty ideal. In G. D. Johnson, K. D. Thomas, A. K. Harrison, & S. A. Grier (Eds.), *Race in the marketplace.* (pp. 73–85). Palgrave Macmillan.

Zimmer, M. (2010). 'But the data is already public': On the ethics of research in Facebook. *Ethics and Information Technology,* 12, 313–325.

Zug, M. A. (2016). *Buying a bride: An engaging history of mail-order matches.* New York University Press.

Index

Subjects

Authors